Bread Staling

CRC Series in CONTEMPORARY FOOD SCIENCE

Fergus M. Clydesdale, Series Editor
University of Massachusetts, Amherst

Published Titles:

America's Foods Health Messages and Claims: Scientific, Regulatory, and Legal Issues
James E. Tillotson

New Food Product Development: From Concept to Marketplace
Gordon W. Fuller

Food Properties Handbook
Shafiur Rahman

Aseptic Processing and Packaging of Foods: Food Industry Perspectives
Jarius David, V. R. Carlson, and Ralph Graves

The Food Chemistry Laboratory: A Manual for Experimental Foods, Dietetics, and Food Scientists
Connie Weaver

Handbook of Food Spoilage Yeasts
Tibor Deak and Larry R. Beauchat

Food Emulsions: Principles, Practice, and Techniques
David Julian McClements

Getting the Most Out of Your Consultant: A Guide to Selection Through Implementation
Gordon W. Fuller

Antioxidant Status, Diet, Nutrition, and Health
Andreas M. Papas

Food Shelf Life Stability
N.A. Michael Eskin and David S. Robinson

Bread Staling
Pavinee Chinachoti and Yael Vodovotz

Bread Staling

Edited by
Pavinee Chinachoti
Yael Vodovotz

CRC Press
Taylor & Francis Group
Boca Raton London New York

CRC Press is an imprint of the
Taylor & Francis Group, an **informa** business

First published 2001 by CRC Press
Taylor & Francis Group
6000 Broken Sound Parkway NW, Suite 300
Boca Raton, FL 33487-2742

Reissued 2018 by CRC Press

Library of Congress Cataloging-in-Publication Data
Bread Staling / edited by Pavinee Chinachoti and Yael Vodovotz.
p. cm. -- (Contemporary food science series)
Includes bibliographical references and index.
ISBN 0-8493-8790-6 (alk. paper)
1. Bread -- Analysis. 2. Food spoilage. I. Chinachoti, Pavinee. II. Vodovotz, Yael. III. Series.

TX769 .B77534 2000
664'.7523--dc21 00-059891

A Library of Congress record exists under LC control number: 00059891

Publisher's Note
The publisher has gone to great lengths to ensure the quality of this reprint but points out that some imperfections in the original copies may be apparent.

Disclaimer
The publisher has made every effort to trace copyright holders and welcomes correspondence from those they have been unable to contact.

ISBN 13: 978-1-315-89124-8 (hbk)
ISBN 13: 978-1-351-07034-8 (ebk)

Visit the Taylor & Francis Web site at http://www.taylorandfrancis.com and the
CRC Press Web site at http://www.crcpress.com

Introduction

Bread staling has been a significant problem in the food industry since ancient times. Consumers demand fresh baked goods that do not stale within a reasonable time frame, while still delivering the taste and texture expected from such products. Formulation and processing technologies designed to control the staling rate have long been investigated. Nevertheless, bread remains a processed food with one of the shortest shelf lives. Mold growth, loss of flavor, and rheological changes are common signs of staling, yet the molecular and structural origins of these are not clearly defined. The roles of starch, gluten, lipids, water, and other components in bread staling are continuously studied with advanced analytical methodologies and sophisticated multidisciplinary approaches.

Significant progress has been made recently in the fundamental understanding of events leading to staling of bread. This book presents the current state of knowledge of the mechanism of bread staling from a physiochemical perspective. Various research groups are currently investigating the mechanism of staling, and much of their recent work is included in our book. Dr. Schiraldi and Dr. Fessas elegantly present an overview of the current understanding of bread staling in Chapter 1. This chapter also contains some of their new work utilizing pentosans, which may lead to a significant delay of bread firming. This overview is followed in Chapter 2 by Dr. Cesàro's in-depth look at material science as it applies to bread biopolymers, which serves as the basis for understanding complex biopolymer system behavior in bread. Dr. Parker and Dr. Ring elaborate in Chapter 3 on the role additives play in delaying bread staling. Chapters 4 and 5 introduce the reader to the numerous techniques used in bread staling research. Special care is taken to describe the benefits and limitations of each technique. Chapter 6 is dedicated to the advancements of nuclear magnetic resonance and the great promise it holds for understanding the physiochemical properties of bread components. In Chapter 7, Dr. Le Meste and co-authors describe the behavior of starch upon heating, and particularly, the role of water. Dr. Oates summarizes the current understanding of bread microstructure in Chapter 8. The concluding chapter by Dr. Farhat and Dr. Blanshard describes a way to model starch crystallization, a key factor in the firming of bread.

The fundamental events leading to bread staling, with the intent of ultimately improving existing shelf life and designing new and longer lasting baked goods are presented in this book. The use of material research and molecular spectroscopy to solve a food problem is a new way to approach the centuries-old dilemma. This approach will lead to further fundamental understanding and will aid manufacturers in the future development of anti-staling formulations for bread and other bakery products.

Numerous technological solutions have been proposed to solve bread staling problems. Many of them work and many do not work under industrial conditions. Such inconsistent outcomes may lie in the serious lack of adequate fundamental knowledge of molecular structure and function relationships in such a complex system. We hope that experts' viewpoints drawn together here will be beneficial to fundamental researchers and technologists who are interested in understanding the complex mechanisms underlying bread staling problems. The editors graciously thank all contributing authors involved in this book.

Pavinee Chinachoti
Yael Vodovotz

Editors

Pavinee Chinachoti earned her B.S. degree in biology at Mahidol University (Thailand), and her M.S. and Ph.D. in food science at the University of Illinois at Urbana-Champaign. She then became a faculty member at the University of Massachusetts (Amherst). She is a full professor, teaching and conducting research in physiochemical properties in foods, with a specialization on the role of water in food.

Dr. Chinachoti develops and directs research in control of water interaction and migration to improve food product shelf life stability. Her goal is to investigate the role of water to control physical, chemical, and microbial changes in food for improved stability in quality and safety. Emphasis is on value-added technology related to starch, gluten, and sugars, applied to food products such as bread and other intermediate moisture and high moisture foods to prolong shelf-life. Recent work has focused on molecular dynamics, using nuclear magnetic resonance (NMR) and thermal analysis to investigate molecular motions in foods as related to food stability.

Dr. Chinachoti teaches food processing and water in foods courses. She has been selected as a Lilly Teaching Fellow and received the College of Food and Natural Resources (CFNR) Outstanding Advisor Award (1998), Eastern Food Conference Outstanding Professor Award (1999), and CFNR Certificate of Excellence in Advising.

Dr. Chinachoti is active in the Institute of Food Technologists (IFT) where she has served as a member and chair of the Committee on Education, chair of Long-Range Planning Committee for the Food Chemistry Division, past jury member and chair for IFT scholarship program, and active organizing member of the Carbohydrate Division. She has been elected eastern director for the Association of Thai Professionals in America and Canada (ATPAC) and is a team leader, bringing U.S. teams to Thailand to assist public universities in autonomous reform in collaboration with the Ministry of University Affairs of Thailand.

She has published more than 70 papers and over 50 abstracts, and she has made more than 60 presentations in the past ten years.

Yael Vodovotz earned her B.S. in food science from the University of Illinois at Urbana-Champaign, her M.Sc. from the University of British Columbia, Vancouver, and her Ph.D. from the University of Massachusetts (Amherst). She joined NASA-Johnson Space Center as a postdoctoral fellow and held a managerial post at the Advanced Life Support Food System for one year. She is an assistant professor at Ohio State University, teaching and conducting research in physiochemical properties in foods, with emphasis on baked products and their components.

Dr. Vodovotz's research at Ohio State University is in the area of carbohydrate chemistry, with focus on water mobility and stability in starch-based products, and development of baked goods with extended shelf life. She plans to collaborate with plant researchers to explore changes in starch functionality in plants grown in little or no gravity, development of food from inedible byproducts of hydroponically grown plants, and time-release systems such as drug delivery and flavor-release mechanisms. Her teaching includes cereal chemistry and advanced food analysis methods.

Dr. Vodovotz is assistant editor for the *Journal of Life Support* and *Biosphere Science*. She served on an expert panel for NASA research announcement grant review, and represented NASA-Johnson Space Center at the conference of the International Committee for Material Circulation in Geo-Hydrosphere and its Application in Rokkasho, Japan. Dr. Vodovotz received the ACS Agriculture and Food Chemistry Division Withycombe fellowship and the American Association of Cereal Chemists graduate fellowship.

She has published more than 20 papers and made over 30 presentations in the past ten years.

Contributors

Mooyeol Baik
Department of Food Science
University of Massachusetts
Amherst, MA

J.M.V. Blanshard
Division of Food Sciences
School of Biosciences
University of Nottingham
Loughborough, United Kingdom

Silvana Cavella
Dipartimento di Scienza degli Alimenti
Universita Degli Studi di Napoli Federico II
80055 Portici (NA)
Italy

Attilio Cesàro
Laboratory of Macromolecular Chemistry
Department of Biochemistry, Biophysics, and Macromolecular Chemistry
University of Trieste
Trieste, Italy

Paul L. Chen
Department of Biosystems and Agricultural Engineering
University of Minnesota
St. Paul, MN

Pavinee Chinachoti
Department of Food Science
University of Massachusetts
Amherst, MA

Eleni Chiotelli
Equipe d'Ingénierie Moléculaire et Sensorielle de l'Aliment
Ecole Nationale Supérieure de Biologie Appliquée à la Nutrition et à l'Alimentation
Université de Bourgogne
Dijon, France

Imad A. Farhat
Division of Food Sciences
School of Biosciences
University of Nottingham
Loughborough, United Kingdom

Dimitrios Fessas
Dipartimento di Scienze e Tecnologie Alimentari e Microbiologiche
Università degli Studi di Milano
Milano, Italy

Martine Le Meste
Equipe d'Ingénierie Moléculaire et Sensorielle de l'Aliment
Ecole Nationale Supérieure de Biologie Appliquée à la Nutrition et à l'Alimentation
Université de Bourgogne
Dijon, France

Paolo Masi
Dipartimento di Scienza degli Alimenti
Universita Degli Studi di Napoli Federico II
80055 Portici (NA)
Italy

Christopher G. Oates
Agro Food Resources Co. Ltd.
Nonthaburi, Thailand

Roger Parker
Food Biopolymer Section
Institute of Food Research
Norwich Research Park
Norwich, United Kingdom

Laura Piazza
Dipartimento di Scienze e Tecnologie Alimentari e Microbiologiche
Università degli Studi di Milano
Milano, Italy

Stephen G. Ring
Food Biopolymer Section
Institute of Food Research
Norwich Research Park
Norwich, United Kingdom

Arnaud Rolée
Equipe d'Ingénierie Moléculaire et Sensorielle de l'Aliment
Ecole Nationale Supérieure de Biologie Appliquée à la Nutrition et à l'Alimentation
Université de Bourgogne
Dijon, France

R. Roger Ruan
Department of Biosystems and Agricultural Engineering
Department of Food Science and Nutrition
University of Minnesota
St. Paul, MN

Alberto Schiraldi
Dipartimento di Scienze e Tecnologie Alimentari e Microbiologiche
Università degli Studi di Milano
Milano, Italy

Fabiana Sussich
Laboratory of Macromolecular Chemistry
Department of Biochemistry Biophysics and Macromolecular Chemistry
University of Trieste
Trieste, Italy

Elena Vittadini
NASA-Johnson Space Center
Houston, TX

Yael Vodovotz
Department of Food Science and Technology
The Ohio State University
Columbus, OH

Table of Contents

1 Mechanism of Staling: An Overview

Alberto Schiraldi and Dimitrios Fessas

CONTENTS

INTRODUCTION

Freshly baked bread can be easily recognized by its sensory attributes: crisp crust, soft crumb, and appealing aroma. The starch granules of the crumb are swollen and deformed but still recognizable on microscopic inspection, because their swelling is limited by the deficiency of water. Starch granules are separated from each other by a protein layer (~1 μm thickness) which looks like a continuous phase throughout the crumb.[1,2] The granules in the crust are less swollen, displaying some birefringence under polarized light.[3] During aging, crumb firmness significantly increases, crispness of the bread crust decreases, and the bread loaf loses its fragrance, assuming a stale flavor. Concurrently, some fractures appear in the continuous phase in which the remnants of starch granules seem separated from the surrounding protein matrix.[2] The bread-making process including the dough recipe, the method of mixing and proofing, and the thermal history of the dough during baking, can significantly affect staling.

Bread can be referred to as an unstable material, and bread staling as a multifaceted process involving physical, chemical, and sensory changes that are related.

Bread staling has been the subject of a number of studies to obtain an understanding of its intrinsic mechanism and thereby modify dough recipes to increase shelf life.[4-11] Because of the complexity of the process, most studies address specific bread properties during its shelf life. An overall description of the staling process remains to be evaluated.

0-8493-8790-6/01/$0.00+$.50

Studies of the increase of crumb firmness are more prevalent in the literature than those devoted to the other attributes of stale bread. Most experimental approaches have involved x-ray diffraction, differential scanning calorimetry (DSC), and rheological investigations. These techniques were complementary to one another as long as they described relations between microscopic structure and macroscopic behavior of the product. A number of models have been proposed[4,9,12,13] to describe the increase of firmness of the bread crumb as well as the possible anti-firming roles of surfactants, non-starch polysaccharides, and enzyme treatments. However, the relevant conclusions have left some major issues unexplained, such as the role of gluten in the firming process.

A general consensus is that water plays a critical role in the firming process, either by enhancing the molecular mobility of polymer chains or by acting as a coordination agent between them. A new investigative approach has been introduced to establish the role of short and long range molecular mobility. This approach requires selective use of spectroscopic techniques such as ^{1}H- and ^{17}O NMR,[14-16] cross-polarization magic-angle spinning (CPMAS), Nuclear Magnetic Resonance (NMR),[17] and probe electron spin resonance (ESR).[18] Dynamic mechanical analyses, e.g., thermal mechanical analysis (TMA),[19] dynamic mechanical analysis (DMA),[20] and other experimental approaches were included in the armory of techniques in order to detect relaxation effects in a wide temperature range.

The present chapter cannot replace the excellent book edited by Hebeda and Zobel[21] in which most of the previous studies on bread staling are accurately referenced and discussed. In this chapter we report issues that clarify the mechanism of staling and suggest some areas for future investigations.

ROLE OF STARCH AND OTHER MAIN CRUMB COMPONENTS

Starch is significantly involved in the staling process. Its main transformation upon aging, *retrogradation*, is the aggregation of polysaccharide chains which may form crystal phases within and outside the contours of the native starch granules. A moderate fraction (15 to 30%)[22] of both amylose and amylopectin of gelatinized bread starch can undergo this change, whereas the rest of the starch remains amorphous.

The starch chains can assume single and double helix conformations. In crystalline phases these helices are regularly displaced around a hexagonal axis.[22] A molecule of amylose or amylopectin can exist in segments arranged in a helical conformation as well as in unordered segments, contributing to the amorphous fraction of the starch. The crystalline phases whose patterns are recognized from X-ray diffraction usually are classified as A, B, C, and V pattern.[23] The V phase is formed by amylose and amylopectin single helices trapping or interacting with a lipid molecule. The hydrophobic tail of the lipid is extended along the internal cavity of the helix, while its hydrophilic head protrudes from the end of the coil.[24] The B phase is caused by double-helix amylose chains and amylopectin side chains and generally includes 27% water. The A and C phases, which tend to contain less than 27% water, are formed at higher temperatures.

Many of the staling mechanisms proposed so far have been accounted for and, to some extent, integrated in hypothetical models such as the one proposed by Zobel and Kulp.[13] This model explains the firmness of stale crumb as originating from the formation of tight junctions between polysaccharide chains, which may eventually progress toward the formation of crystalline structures (mainly B and V types). These junctions are primarily the result of inter-chain hydrogen bonds. Formation of double helices, involving either segments of amylose chains or side moieties of amylopectin molecules which did not arrange in any crystal order, contributes significantly to an increase in crumb firmness. The process can take place either inside or outside the starch granules. Accordingly, some additives (e.g., surfactants and lipids), which can reduce inter–granule cohesion (by reducing leaching of amylose upon formation of amylose-lipid complexes) and hardness of granules (lipids and simple sugars migrating into the granules), would lead to a softer crumb.

This model regards water mainly as a plasticizer of the inter-granule amorphous matter and as a medium where crystalline phases can grow more rapidly, since polymer chains can move more easily with respect to each other. During storage of bread, some water migrates from amorphous to crystalline starch, where it is more tightly bound, resulting in an increase of overall crumb firmness and hardness. Gluten may not play any significant role, forming a continuous phase which bridges the surfaces of neighboring granules.

Although this model is consistent, it neglects some major points. Starch and gluten would undergo changes during staling, and some water redistribution would take place from one component to another within the dough[5] and the crumb.[7] According to Kulp and Ponte,[7] changes in starch may cause staling in the early storage period, while gluten would produce effects in the latter stages.

The role of proofing, which largely contributes to the alveolar structure of the crumb and therefore to the value of its elastic modulus,[25] is another important issue which is often neglected. This was mentioned in the review by Zobel and Kulp[13] but not included in their model. Compliance of the gluten network and the role of water-soluble proteins, oligosaccharides, and lipids is also relevant to this phenomenon.[25-28] Once compressed to expel all the air phase, bread crumbs show the same elastic modulus regardless of the recipe used.[28]

A correlation between crumb firmness and starch crystal fraction has been found by many authors, but some aspects still require further evidence. For example, there is no model illustrating that amylose crystallization (which is not reversible by a refreshing treatment, since the refreshing temperatures are below the melting points of V- and B-crystals) does not play a relevant role in firming of crumb. Additionally, why is an increase in crumb firmness often accomplished by an increase in crystal fraction of amylopectin? This is not always true for breads prepared from the same flour with minor composition differences, such as additional water, gluten, or soluble proteins.[29] The same is true for some anti-staling agents, like pentosans, which do not equally affect the extent of starch retrogradation determined by DSC.[28,30]

The extent and the rate of starch retrogradation evaluated by DSC or by dynamometric means (a typical instrument is the Instron Universal Testing Machine) are greater than those assessed via x-ray diffraction. DSC is more sensitive to the

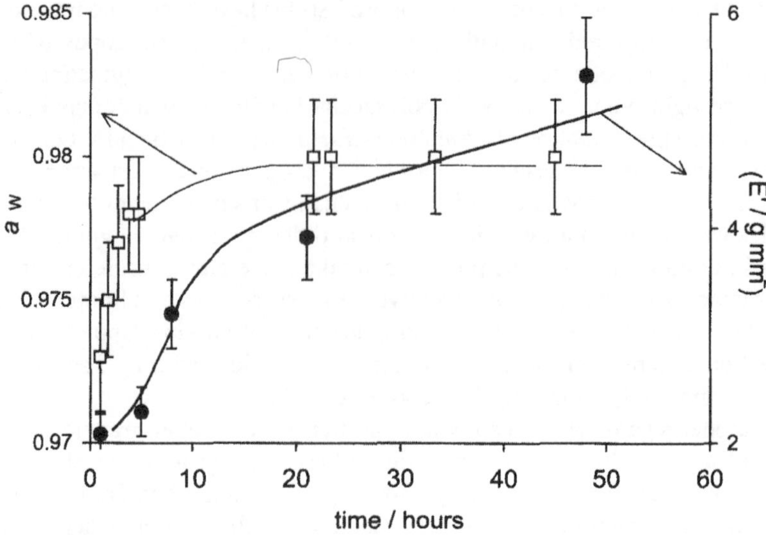

FIGURE 1 Change of elastic modulus, E', and water activity, a_W, of bread crumb in the early staling phases at room temperature.[29]

formation of bonds that may not imply growth of crystal phases, but can affect the increase of crumb firmness.[31] However, this could be accounted for if other phenomena not related to starch play a significant role in crumb firming.

According to Willhoft,[5] crumb firmness is related to starch retrogradation (referred to as growth of crystal phases) as well as other processes not suggested by the author which play a significant role. The rates of these processes may increase with rising temperature. The starch retrogradation rate, which is at its maximum at about 5°C, declines with increasing temperature. As described in the model by Martin et al.,[12] gluten forms hydrogen bonds with some partially leached amylose and/or amylopectin molecules protruding into the extra-granule region. No experimental evidence of a chemical interaction between starch and gluten was provided, yet this model could be used to explain some rheological phenomena, such as crumb firming in microwave-refreshed breads.[32] However, it does not correlate with the evidence that gluten-rich crumbs, e.g., from durum wheat flours, undergo a slower staling.[33,34]

The change of water activity (a_W) which takes place in bread crumb is an important issue. Water activity decreases with crumb aging and parallels that of the corresponding water content.[35] In the first 10 hours of aging,[29] an opposite trend was observed (Figure 1), followed by a plateau that was also recognized in NMR investigations.[36] It was also found that water activity in a 24 hour-aged crumb was greater than that in the corresponding starting dough.[35] Larger a_W means larger fugacity, i.e., tendency to escape. In other words, water would be freer to move in freshly baked crumb than in the dough, and even freer during the early phases of bread staling.

The following questions should be addressed:

- Which water is detected in a_W determinations in a viscous system such as bread crumb? It seems reasonable to speculate that it could be the most mobile fraction. According to some NMR investigations, this should be the water within the amorphous regions.[16,37]
- How reliable are a_W determinations in systems when measured at temperatures above their own T_g?[38] It is easy to argue that the "experimental" a_W value might be lower than the actual one which would require some equilibration. In the case of crumb staling, this could imply a decrease of the observed a_W with bread aging related to the decrease of crumb T_g.[19]

According to ^{17}O NMR investigations,[16] some water is actually bound, although the binding substrates cannot be distinguished, and analogous conclusions can be drawn from FTIR investigations.[39] Taking into account that bread crumb is not a true thermodynamic equilibrium, water is expected to move from regions where its chemical potential is higher to regions where chemical potential is lower, i.e., from a less-bound to more-bound state. This would, accordingly, imply a decrease of a_W with bread aging.

The finding of an a_W increase in the earliest phases of bread staling might instead correspond to some initial modifications within the crumb. Therefore, water behavior in staling bread crumb might depend not only on gradients of its chemical potential, but also on other driving forces related to the surrounding matrix.[29] As previously mentioned, newly developed experimental and theoretical approaches will allow a more direct investigation of changes at the molecular level. Some of these methods are reported in this book to be used by those who want to contribute to knowledge in this subject.

ROLE OF ANTI-STALING COMPOUNDS: WHEN AND WHERE THEY ARE SUPPOSED TO ACT

Lipids and Surfactants

Lipids affect loaf volume and produce a significant anti-staling effect.[26,40-42] A number of mechanisms of their action have been suggested. It has been clearly established that fatty acids, monoglycerides, and diglycerides can be included within the central cavity of amylose and, to a lesser extent, amylopectin single helices,[43] therefore enhancing the formation of the V-type crystals.[22,44] Similar complexes were also revealed by means of synchrotron x-ray diffraction.[45] These complexes, however, did not have a specific role in crumb firming.[46] Other explanations were therefore proposed which relate the role of softening or anti-firming additives to their effect on loaf volume.[5]

Some years ago Willhoft[5] suggested that the anti-staling effect produced by monoglycerides (especially unsaturated) could result from an interaction with gluten, which has since been confirmed.[47-49] Surfactants are trapped by the gluten phase during dough mixing, then released toward the starch gel during baking. The surfactants would remain primarily in the intergranule regions[50] where they can form complexes with leached amylose and/or amylopectin. This can inhibit amylose and amylopectin from forming bridges between swollen starch granules and ultimately prevents retrogradation.

NON-STARCH POLYSACCHARIDES

Non-starch polysaccharides, like arabinoxylans and arabinogalactans, which are often bound to a protein moiety, are also believed to have an anti-staling effect.[51] These polysaccharides, currently referred to as pentosans, are native components of cereal flours which can be separated from starch and gluten by means of water-ethanol extractions.[52] The water-extractable pentosans, which have relatively small molecular mass (20 to 70 kDa), are believed to be less effective than larger homologues at delaying staling.[53] Addition of small amounts of pentosans (0.5 to 2%) to a standard bread dough recipe may produce a moderate increase of loaf volume[51] (a controversial point),[54] and a significant decrease of the firming rate of the final product.[51] A number of tentative interpretations of the relevant anti-staling mechanism have been reported,[51-56] including possible interaction with gluten.[57,58] It has been clearly assessed that these compounds do not affect starch retrogradation,[59,28] although they produce a delay of crumb firming.[60] This also depends on the pentosan molecular mass, which can be enzymatically controlled.[61-63] A critical review of these results is not given here. A specific research project supported by the European Commission to establish the properties of purified arabinoxylans and their role in breadmaking and bread staling is in progress.[64] The current belief[57] is that pentosans could act as molecular water sinks within the crumb matrix, where they would slowly release moisture in the course of bread shelf life. Because of its plasticizing effect, the water released would keep the crumb softer for a longer time span. This interpretation is based on the finding that pentosans could retain large amounts of water.[51] It is possible that some imbibing water can be physically trapped within a pentosan gel, but the actual hydration of these polysaccharides is comparable to that observed in an equivalent mass of xylose or arabinose (Figure 2). Complete hydration of dry pentosan cotton-like flakes requires several hours under continuous stirring,[53] implying poor rate of moisture uptake by native flour pentosans during dough mixing. No reliable correlation was found between increased dough viscosity and improved loaf volume or anti-staling effect of these compounds.[53]

According to some recent results, the alveolar crumb structure deserves more attention than the overall loaf volume. The alveolar structure of bread crumb produced from pentosan-enriched dough was found to be coarser than that in the control bread. The action of pentosans could therefore be focused on the formation of dough and crumb structure, which would affect the compliance of the matrix around the gas cells formed during proofing and early baking. The main effect of pentosans might, therefore, be concentrated on the immediate environment of the gas/bulk inter-phase. The alveolar crumb structure plays an important role in the sensory texture of a bread. The coarser the crumb structure, the softer the crumb appears and the slower the firming rate with aging, irrespective of starch retrogradation and moisture loss.[28]

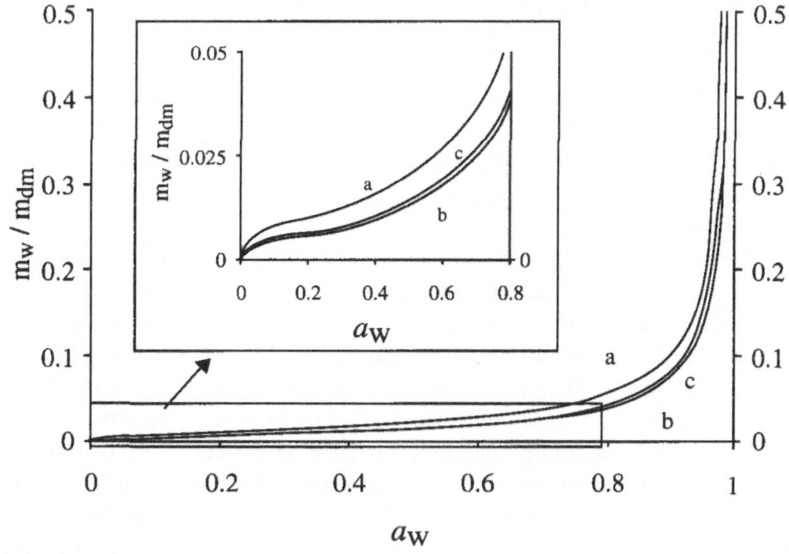

FIGURE 2 Desorption isotherm of (a) low molecular weight arabinoxylans, (b) arabinose, and (c) xylose. m_W and m_{dm} stand for mass of water and mass of dry matter, respectively.[65]

Bread can be prepared from a flour enriched with small amounts of water-soluble flour proteins[66] extracted from a separate lot of the same flour. In this case, the bread's alveolar crumb structure is finer and its firmness increased with respect to the control bread at any aging or storage time.[28] This finding suggests that the effect of pentosans might be counterbalanced by that of water-soluble flour proteins, referred to as albumins and globulins,[66] which are believed to act as surfactants within a thin liquid film layering the gas cell walls.[67,68]

BREAD FLAVOR

As stated by Willhoft,[5] one should be aware that, "although the firming of the crumb is the simplest and the most important issue of bread staling, a model depicting the staling process should also account for the other phenomena, namely, deterioration of flavor and loss of crust crispness."

Baking results in browning reactions[4] with formation of volatiles (alcohols, aldehydes, diacetyl, ketones [mainly 2-heptanone], esters, etc.) which disappear during aging and are partially replaced by other degradation products responsible for the characteristic stale flavor. The volatiles do not disappear through volatilization but instead are locked within the amylose helices which prevent them from being perceived to the taste. They are partially released on refreshing the bread.[5] Taking into account the role of water described above, it seems reasonable to speculate that flavor retention should also be related to the migration of the solvent. However, no detailed study has been reported, although the subject has been discussed.[69] According to Reineccius,[70] more research efforts are needed in this field.

BREAD AS A DISPERSED SYSTEM: A PERSPECTIVE TO DESCRIBE BREAD STALING

Gluten, soluble proteins, and pentosans are responsible for adsorbing much water within a mixed dough. When dough is heated during baking, some water is involved in starch gelatinization but a significant fraction of it remains attached to the other polymers and solutes, which compete for the available moisture. This competition produces phase separation.

By ultra-centrifuging dough samples, Larsson and Eliasson[71] were able to separate four aqueous phases, namely, (1) a liquid phase that contained dispersed starch granules, sugars, and soluble proteins, (2) a gel phase that contained large amounts of damaged starch and pentosans, (3) a gluten phase (with some starch granules), and (4) a starch phase. Taking into account that centrifugation is indeed a way to accelerate the change of finely dispersed systems into bulk separated phases, one may conclude that bread dough can be referred to as a metastable dispersed system with a large surface across which water can move from one phase to another.

The main parameter governing this phase separation is the difference between excluded volumes of polymer components of the dough. The resulting immiscibility reflects the thermodynamic incompatibility between different polymers such as proteins and polysaccharides.[72] In relatively concentrated systems, thermodynamic incompatibility produces separate aqueous phases of minute geometrical size (1 to 5 μm). In these water-in-water emulsions, water exchanges are governed by osmotic pressure[73] occurring across gradients of water chemical potential.

The formation of a 3D network inhibits the attainment of a thermodynamic equilibrium between coexisting phases.[74] Local viscosity can significantly affect water repartition or redistribution, which takes a relatively long time. Mechanical treatment can thus affect the water-binding capacity of dough and water partition among various dough phases.[75,76] The gluten phase, which is a thixotropic gel, has a water content that may be increased by overmixing the original dough without increasing the overall dough water content.[71] Overmixed dough samples are more sticky despite the fact that gluten is expected to become weaker. This behavior may indicate a change in the gluten structure such as that caused by cross-linking.[71]

Gliadin and glutenin fractions of wheat flour are not miscible with albumins, globulins, soluble starch, or pentosans.[72] Aggregation of gliadins, which takes place at a lower temperature when their concentration is increased, is anticipated in the presence of arabinoxylans (Figure 3).[64]

These results can be interpreted to mean that the starting mixture of gliadins and pentosans would split into two phases, a gliadin-rich phase and a pentosan-rich phase, as usually observed in aqueous systems of thermodynamically incompatible polymers (Figure 4).

Further evidence of separation of the liquid phase, observed by Larsson and Eliasson,[71] was concurrent with the expectations of thermodynamic incompatibility between starch polysaccharides and pentosans[74] and possibly between these and flour-soluble proteins.[64] Finally, as a reminder, amylose and amylopectin in aqueous medium are also incompatible.[78] The real number of separated and dispersed phases

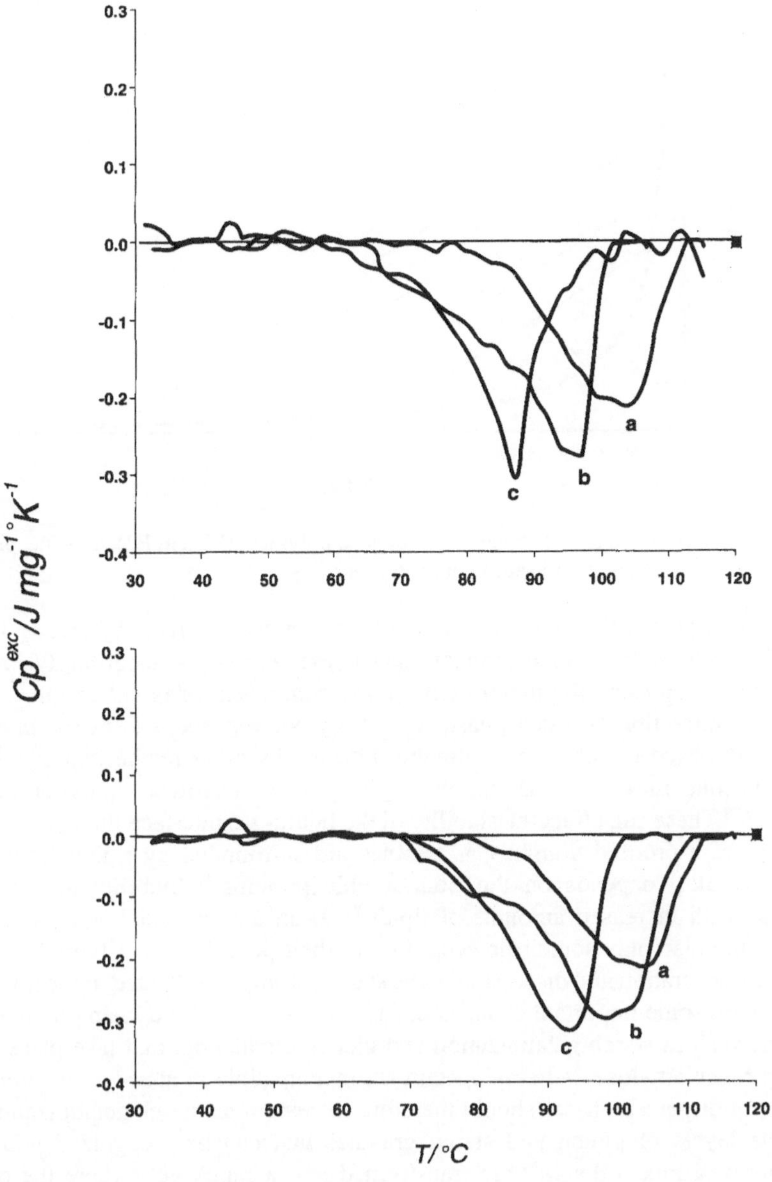

FIGURE 3 DSC traces[64] showing the exothermic signal from aggregation of gliadins in aqueous solution. The upper figure shows that aggregation occurs at lower temperature when the protein concentration is increased (a = 25, b = 50, c = 75 mg mL^{-1}). The lower figure shows an analogous effect produced by addition of arabinoxylans to a 25 mg mL^{-1} gliadin solution (a = 0, b = 26.4, c = 66 mg mL^{-1}).

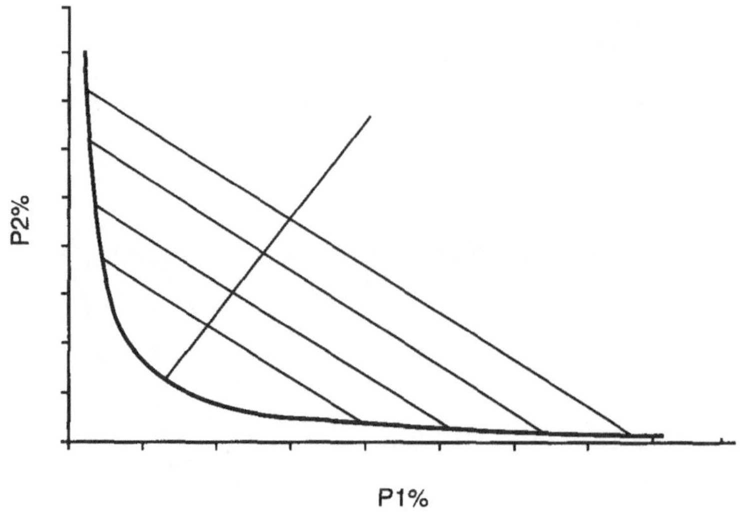

FIGURE 4 Phase separation between two aqueous polymers (P1 and P2). Tie-lines intercept the binodal curve at the equilibrium concentrations.

within a dough may therefore exceed that of the phases separated by means of ultra-centrifugation. A large extension of inter-phase regions is accordingly expected which might represent the major contributing component of bread crumb.

Water migration between phases of largely heterogeneous systems, like bread dough and baked crumb, can be directly affected by other recipe ingredients with low molecular mass, such as simple sugars, oligosaccharides, lipids, and surfactants.[75,76,78] These can affect the stability of the liquid-gas interface lining the alveolar structure of a proofed dough. Gas bubbles are surrounded by a thin liquid layer whose stability depends on the flour-soluble proteins.[78] Stability also tends to decrease with increased amounts of lipids.[79] As in any colloidal aqueous system, salts should also play some role according to their position in the lyotropic series.

Phase separation and dispersion in the starting dough is reflected in the final bread crumb, with some important changes due to modification of the main polymer components, such as starch gelatinization and gluten reticulation, that take place during baking. Since amylose and amylopectin are incompatible in aqueous medium,[80] the number of dispersed phases should therefore increase after starch gelatinization. The alternate layers of gluten and starch granules that can be recognized within the structure of a mixed dough[71] are transformed into a starch gel (where the original contours of granules can be still recognized[5]) inter-penetrated[75] by a thermoset gluten network. The remnants of the liquid layers surrounding the gas bubbles are regions where denatured flour-soluble proteins are more concentrated.[67,68]

Much water is still in the dough. Its high thermodynamic activity (above 95%[29,35]) suggests that it should be somewhat free to move along any gradient of the relevant chemical potential throughout the crumb structure. Any temperature rise, aside from its effects such as melting of retrograded amylopectin, would enhance water displacement between the coexisting phases and in turn modify the relevant glass transition temperature, T_g, and the rate of many relaxation processes.

Some features of bread staling that are difficult to explain with starch retrogradation only can be interpreted in a more general picture. When polymers are thermodynamically incompatible they can use the water available within their own phase to sustain conformational and phase transitions, depending on its local activity and concentration. It is therefore clear, for example, why some water is necessary to promote starch retrogradation.[10,35] If the phase considered has a high viscosity, at a temperature below its own T_g, no major physical processes can take place during a time span of a few days.

According to Willhoft,[5] molecular changes take place in both gluten and starch. Molecular changes in the gluten fraction may be similar to those described by Muller,[81] i.e., leading to the formation of cross-links. The formation of such cross-links, either in gluten or in starch gel, could be associated with the production of free water, i.e., release of water originally bound to the polymer chain.[29] Bushuk and Hlynka[83] have indicated that the 3D protein structure in dough is held together by H-bonds, disulphide links, and salt links, and by H bonds involving water. When cross-links are formed between adjacent polymer chains, water is expelled. This can facilitate the formation of new cross-links keeping the relevant binding sites closer.[29,83] Heating of stale bread accordingly produces irreversible modifications of the gluten structure, unlike the well-known heat-reversible character of retrograded starch. An irreversible step of thermoset process involving water release has been included in a tentative model.[29]

As a result of the transformation of polymer components, water can be more tightly trapped within a given polymer phase, where some water molecules remain bound within the crystal structure, or more easily released toward phases containing other polymers, possibly in the state of viscous gels.

It has therefore been suggested[83,84] that free and bound water could play different roles in determining the staling rate, namely, a rapid staling rate below 30°C and a much slower rate above 30°C. This would be the temperature at which a change of the state of water could take place. Similar evidence came from thermogravimetry (TGA) investigations on staling bread crumb.[29] In these investigations, water release was divided into two main contributions, one with a maximum rate of about 70°C and one at 90°C (Figure 5). These went through a minimum and a maximum rate, respectively, at about 10 hour aging, which related to a maximum in the crumb elastic modulus as well as a maximum water activity value.

The release of structurally-bound water by hydrated gluten during baking, and then at a slower rate during storage at room temperature, has been discussed.[85] One can accordingly accept some of Willhoft's hypotheses[5]:

- Starch would increase its degree of hydration and retrogradation by acting as a sink for mobile moisture
- Crumbliness of stale bread could depend on the partial dehydration of gluten

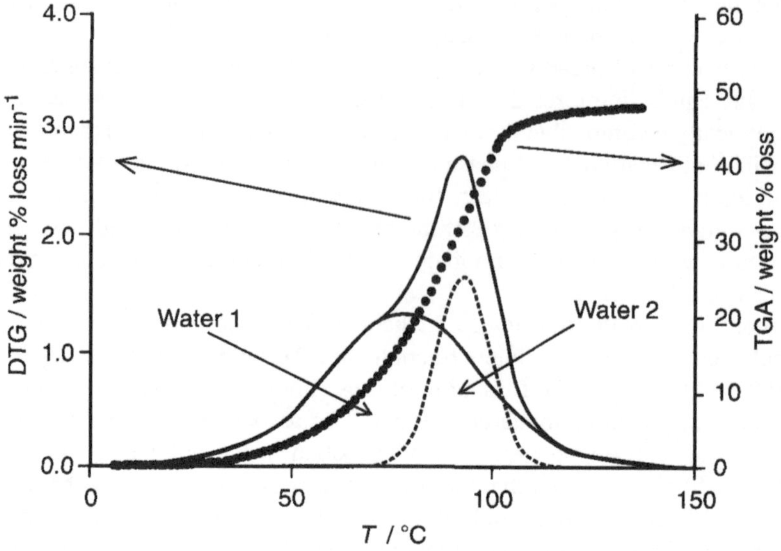

FIGURE 5 Thermogravimetric analysis (TGA) and its time derivative (DTG) traces of 5 hour aged bread crumb. Two Gaussian functions were used to split the DTG trace into two contributions.[29]

- An increase of gluten/starch ratio produces greater resilience and smoothness of the crumb
- Crumb harshness observed in bread with low gluten levels would accordingly be due to the loss of moisture from gluten either toward starch or to the crust region, or possibly both.

The overall picture of the crumb could be described as interpenetrated gels separated by aqueous inter-phases which contain most of the low molecular weight solutes. This water is rather mobile and can facilitate mutual displacement of the incompatible gel phases, thus behaving as a plasticizer, and can enhance the crumb-to-crust migration of moisture. This local drying makes the walls of the crumb alveoli more rigid, while the concurrent moisture increase within the crust region is accompanied by a reduction of crispness even when overall moisture loss is prevented by packing bread in sealed bags.[86] Along its way toward the crust, water can contribute to a closer packing of the structure through which it is moving, either within a given phase or at the inter-phases, by tightening the sites able to form H bonds. This would explain why refreshed bread softens when its temperature has been raised above T_g, but then becomes harder than the starting staled product, and why microwave-cooked or refreshed bread shows a fast firming without significant enhancement of amylopectin crystallization.

FUTURE ISSUES

The study of inter-phases will be crucial to improve the present knowledge of physics and chemistry of bread and bread staling. Since exchanges of water take place throughout storage, they should directly involve transport processes across the intermediate regions, where many low molecular weight solutes are present together with water soluble proteins. The viscosity of the inter-phases, which is expected to significantly depend on temperature, would therefore play a major role in bread staling and in the refreshing of stale bread. The inter-phase extension is large and is presumably affected by the operative conditions used in bread making, such as duration and strength of mixing, resting, proofing, and thermal history during baking. The same dough recipe can yield significantly different breads when the dough undergoes different treatments. On the other hand, different flours or flour blends may require different bread-making procedures because of their specific compositions.

Therefore, before looking for a magic additive, adjustments in the bread-making process should be considered, recognizing the role of each ingredient and the interactions between ingredients, leading to improved baking performance of the available flours.

REFERENCES

1. Kay, M. and Willhoft, E. M. A., Bread staling. IV. Electrical properties of the crumb during staling, *J. Sci. Food Agric.*, 23, 321, 1972.
2. Mälkki, Y., Paakkanen, J., and Eerola, K., Effect of freezing and monoglycerides on staling of bread, *J. Food Process. Preservation*, 2, 101, 1978.
3. Olsson, H. and Skjöldebrand, C., Microscopic studies on meat products and bread during heat treatment, in *Engineering and Food*, Vol. 1, McKenna, B.M., Ed., Elsevier, London, 1984, 343.
4. Schoch, T. J., Starch in bakery products, *Bakers Dig.*, 39, 48, 1965.
5. Willhoft, E. M. A., Mechanism and theory of staling of bread and baked goods, and associated changes in the textural properties, *J.Texture Stud.*, 4, 292 1973.
6. Maga, J. A., Bread staling, *CRC Crit. Rev. Food Technol.*, 5, 443, 1975.
7. Kulp, K. and Ponte, J. G., Jr., Staling of white pan bread: fundamental causes, *CRC Crit. Rev. Food Sci. Nutr.*, 15, 1, 1981.
8. Russell, P. L., A kinetic study of bread staling by differential scanning calorimetry and compressibility measurements. The effect of added monoglycerides, *J. Cereal Sci.*, 1, 285, 1983.
9. Morris, V. J., Starch gelation and retrogradation, *Trends in Food Sci. Technol.*, July 2, 1990.
10. He, H. and Hoseney, R. C., Changes in bread firmness and moisture during long-term storage, *Cereal Chem.*, 67, 603, 1990.
11. Champenois, Y., Colonna, P., Buléon, A., Della Valle, G., and Renault, A., Gélatinisation et rétrogradation de l'amidon dans le pain de mie, *Science des Aliments*, 15, 593, 1995.
12. Martin, M. L., Zeleznak, K. J., and Hoseney, R. C., A mechanism of bread firming. I. Role of starch swelling, *Cereal Chem.*, 68, 498, 1991.
13. Zobel, H. F. and Kulp, K., in *Baked Goods Freshness*, Hebeda, R. E. and Zobel, H., Eds., Marcel Dekker, New York, 1996, chap. 1.
14. Bulpin, P. V., Welsh, E. J., and Morris, E. R., Physical characterization of amylose-fatty acid complexes in starch granules and in solution, *Starch*, 34, 335, 1982.
15. Wu, J. Y. and Eads, T. M., Evolution of polymer mobility during aging of gelatinized waxy maize starch. ^{1}H NMR study, *Carbohydr. Polym.*, 20, 51, 1993.

16. Kim-Shin, M-S., Mari, F., Rao, P. A., Stengle, T. R., and Chinachoti, P., ^{17}O nuclear magnetic resonance studies of water mobility during bread staling, *J. Agric. Food Chem.*, 39, 1915, 1991.
17. Morgan, K. R., Fourneaux, R. H., and Stanley, R. A., Observation by solid-state 13C CP MAS NMR spectroscopy of the transformations of wheat starch associated with the making and staling of bread, *Carbohydr. Res.*, 235, 15, 1992.
18. Johnson, J. M., Davis, E. A., and Gordon, J., Interactions of starch and sugar water measured by electron spin resonance and differential scanning calorimetry, *Cereal Chem.*, 67, 286, 1990.
19. Le Meste, M., Huang, V. T., Panama, J., Anderson, G., and Lentz, R., Glass transition of bread, *Cereal Foods World*, 37, 264, 1992.
20. Vodovotz, Y., Hallberg, L., and Chinachoti, P., Effect of aging and drying on thermomechanical properties of white bread as characterized by dynamic mechanical analysis (DMA) and differential scanning calorimetry (DSC), *Cereal Chem.*, 73, 264, 1996.
21. Hebeda, R. E. and Zobel, H., *Baked Goods Freshness*, Marcel Dekker, New York, 1996.
22. Zobel, H. F., Starch crystal transformations and their industrial importance, *Starch*, 40, 44, 1988.
23. Katz, J. R., Gelatinization and retrogradation of starch in relation to the problem of bread staling, in *A Comprehensive Survey of Starch Chemistry*, R. P. Walton Ed., Reinhold, New York, 1928, chap. 7.
24. Carlson, T. L. G., Larsson, K., Dinh-Nguyen, N., and Krog, N., A study of the amylose-monoglyceride complex by Raman spectroscopy, *Starch*, 31, 222, 1979.
25. Robertson, G. H. and Emami, S-D., Liquid-jet penetrometry for physical analysis: application to bread aging, *J. Food Sci.*, 39, 1247, 1974.
26. Rogers, D. E., Zeleznak, K. J., Lai, C. S., and Hoseney, R. C., Effect of native lipids, shortening, and bread moisture on bread firming, *Cereal Chem.*, 65, 498, 1988.
27. Wang, Y-J. and Jane, J., Correlation between glass transition temperature and starch retrogradation in the presence of sugars and maltodextrins, *Cereal Chem.*, 71, 527, 1994.
28. Fessas, D. and Schiraldi, A., Texture and staling of wheat bread crumb: effects of water extractable proteins and "pentosans", *Thermochim. Acta*, 323, 17, 1998.
29. Schiraldi, A., Piazza, L., and Riva, M., Bread staling: a calorimetric approach, *Cereal Chem.*, 73, 32, 1996.
30. Piazza, L., Schiraldi, A., Brenna, O., and Vittadini, E., Structure and properties of bread dough and crumb, *J. Therm. Anal.*, 47, 1339, 1996.
31. Roulet, P., MacInnes, W. M., Würsch, P., Sanchez, R. M., and Raemy, A., A comparative study of the retrogradation kinetics of gelatinized wheat starch in gel and powder form using X-rays, differential scanning calorimetry and dynamic mechanical analysis, *Food Hydrocolloids*, 2, 381, 1988.
32. Yamauchi, H., Kaneshigem, H., Fujmura, M., Hashimoto, S., Ohya, K., Hirakawa, T., and Kobayashi, T., Role of starch and gluten on staling of white bread treated with microwave-heating, *Nippon Shokuhin Kogyo Gakkaishi*, 40, 42, 1993.
33. Boyacioglu, M. H. and D'Appolonia, B. L., Characterization and utilization of durum wheat for breadmaking. I. Comparison of chemical, rheological, and baking properties between bread wheat flours and durum wheat flours, *Cereal Chem.*, 71, 34, 1994.
34. Boggini, G., Tusa, P., and Pogna, N. E., Bread making quality of durum wheat genotypes with some novel glutenin subunit compositions, *J. Cereal Sci.*, 22, 105, 1995.

35. Czuchajowska, Z. and Pomeranz, Y., Differential scanning calorimetry, water activity, and moisture contents in crumb center and near-crust zones of bread during storage, *Cereal Chem.*, 66, 305, 1989.
36. Wynne-Jones, S. and Blanshard, J. M. V., Hydration studies of wheat starch amylopectin, amylose gels and bread by proton magnetic resonance, *Carbohydr. Polym.*, 6, 289, 1986.
37. Leung, H. K., Magnuson, J. A., and Bruinsma, B. L., Water binding of wheat flour doughs and breads as studied by deuteron relaxation, *J. Food Sci.*, 48, 95, 1983.
38. Slade, L. and Levine, H., Beyond water activity: recent advances based on an alternative approach to the assessment of food quality and safety, *CRC Crit. Rev. Food Sci. Nutr.*, 30, 115, 1991.
39. Wilson, R. H., Goodfellow, B. J., Belton, P. S., Osborne, B. G., Oliver, G., and Russel, P. L., Comparison of Fourier transform mid infrared spectroscopy and near infrared reflectance spectroscopy with differential scanning calorimetry for the study of the staling of bread, *J. Sci. Food Agric.*, 54, 471, 1991.
40. MacRitchie, F. and Gras, P. W., The role of flour lipids in baking, *Cereal Chem.*, 50, 292, 1973.
41. Russell, P. L., A kinetic study of bread staling by differential scanning calorimetry and compressibility measurements. The effect of added monoglyceride, *J. Cereal Sci.*, 1, 297, 1983.
42. Krog, N., Theoretical aspects of surfactants in relation to their use in breadmaking, *Cereal Chem.*, 58, 158, 1981.
43. Eliasson, A.-C. and Ljunger, G., Interactions between amylopectin and lipid additives during retrogradation in a model system, *J. Sci. Food Agric.*, 44, 353, 1988.
44. Eliasson, A.-C. and Krog, N., Physical properties of amylose-monoglyceride complexes, *J. Cereal Sci.*, 3, 239, 1985.
45. Le Bail, P., Bizot, H., Ollivon, M., Keller, G., Bourgaux, C., and Buléon, A., Monitoring the crystallization of amylose-lipid complexes during maize starch melting by synchrotron X-ray diffraction, *Biopolymers*, 50, 99, 1999.
46. Knightly, W. H., in *Baked Goods Freshness*, R. E. Hebeda and H. Zobel, Eds., Marcel Dekker, New York, 1996, p. 65.
47. Hoseney, R. C., Finney, K. F., Pomeranz, Y., and Shogren, M. D., Functional (breadmaking) and biochemical properties of wheat flour components. V. Role of total extractable lipids, *Cereal Chem.*, 46, 606, 1969.
48. De Stefanis, V. A., Ponte, J. G. Jr., Chung, F. H., and Ruzza, N. A., Binding of crumb softeners and dough strengtheners during breadmaking, *Cereal Chem.*, 54, 13, 1977.
49. Quail, K. J., McMaster, G. J., and Wootton, M., The role of flour lipids in the baking of arabic bread, *J. Cereal Sci.*, 14, 131, 1991.
50. Conde-Petit, B. and Escher, F., Influence of starch-lipid complexation on the ageing behaviour of high concentration starch gels, *Starch*, 46, 172, 1994.
51. Michniewicz, J., Biliaderis, C. G., and Bushuk, W., Effect of added pentosans on some properties of wheat bread, *Food Chem.*, 43, 251, 1992.
52. Cleemput, G., Roels, S.P., Van Oort, M., Grobet, P. J., and Delcour, J. A., Heterogeneity in the structure of water-soluble arabinoxylans in European wheat flours of variable bread-making quality, *Cereal Chem.*, 70, 324, 1993.
53. Courtin, C. M. and Delcour, J. A., Physicochemical and bread-making properties of low molecular weight wheat-derived arabinoxylans, *J.Agric. Food Chem.*, 46, 4066, 1998.
54. Roels, S. P., Cleemput, G., Vandewalle, X., Nys, M., and Delcour, J. A., Bread volume potential of variable-quality flours with constant protein level as determined by factors governing mixing time and baking absorption levels, *Cereal Chem.*, 70, 318, 1993.

55. Biliaderis, C. G., Izydorczyk, M. S., and Rattan, O., Effect of arabinoxylans on bread-making quality of wheat flours, *Food Chem.*, 5, 165, 1995.
56. Izydorczyk, M. S. and Biliaderis, C. G., Cereal arabionoxylans: advances in structure and physicochemical properties, *Carbohydr. Polym.*, 28, 33, 1995.
57. Meuser, F. and Suckow, P., Non-starch polysaccharides, in "*Physics and Chemistry of Baking*", J. M. V. Blanshard, P. J. Frazier, and T. Galliani Eds., Royal Society of Chemistry, London, 1986, p. 42.
58. Saulnier, L., Andersson, R. and Åman, P., A study of the polysaccharide components in gluten, *J. Cereal. Sci.*, 25, 121, 1997.
59. Gudmundsson, M., Eliasson, A.-C., Bengtson, S., and Åman, P., The effects of water soluble arabinoxylans on gelatinization and retrogradation of starch, *Starch,* 43, 5, 1991.
60. Patil, S. K., Finney, K. F., Shogren, M. D., and Tsen, C. C., Water-soluble pentosans of wheat flour. III. Effect of water-soluble pentosans on loaf volume of reconstituted gluten and starch doughs, *Cereal Chem.*, 53, 44, 1976.
61. Cleemput, G., Booij, C., Hessing, M., Gruppen, H., and Delcour, J. A., Solubilisation and changes in molecular weight distribution of arabinoxylans and protein in wheat flours during bread-making, and the effects of endogenous arabinoxylan hydrolysing enzymes, *J. Cereal Sci.,* 26, 55, 1997.
62. Rouau, X., Investigations into the effects of an enzyme preparation for baking on wheat flour dough pentosans, *J. Cereal Sci.,* 18, 145, 1993.
63. Rouau, X. and Moreau, D., Modification of some physicochemical properties of wheat flour pentosans by an enzyme complex recommended for baking, *Cereal Chem.*,70, 626, 1993.
64. Fessas, D. and Schiraldi, A., *Use of Wheat Extractable Arabinoxylans to Improve Stability of Frozen Doughs and Quality of Bread,* unpublished data from FAIR CT97-3069, 1997-2000.
65. Schiraldi, A., Water activity determination by use of TGA with Knudsen cells, in *ISOPOW 7 Book of Abstracts*, Y. H. Roos, Ed., 1998, p. 85.
66. Osborne, T. B., in *The Proteins of the Wheat Kernel*, Publication 84, Carnegie Institute, Washington D.C., 1907.
67. Gan, Z., Angold, R. E., Williams, M. R., Ellis, P. R., Vaughan, J. G., and Galliard, T., The microstructure and gas retention of bread dough, *J. Cereal Sci.,* 12, 15, 1990.
68. MacRitchie, F., The liquid phase of dough and its role in baking, *Cereal Chem.*, 53, 318, 1976.
69. Setser, C. S. in *Baked Goods Freshness*, R. E. Hebeda and H. Zobel, Eds., Marcel Dekker, New York, 1996, chap. 6.
70. Reineccius, G. A., Staling of bakery products, *Cereal Foods World*, 37, 272, 1992.
71. Larsson, H. and Eliasson, A.-C., Phase separation of wheat flour dough studied by ultracentrifugation and stress relaxation, *Cereal Chem.*, 73, 18, 1996.
72. Grinberg, V. Y. and Tolstoguzov, V. B., Thermodynamic incompatibility of proteins and polysaccharides in solutions, *Food Hydrocolloids*, 11, 145, 1997.
73. Tolstoguzov, V. B., Thermodynamic incompatibility of food macromolecules, in *Food Colloids and Polymers: Stability and Mechanical Properties*, E. Dickinson, and P. Walstra, Eds., Royal Society of Chemistry, special publication 113, 1993, 94.
74. Zasyskin, D. V., Braudo, E. E., and Tolstoguzov, V. B., Multicomponent biopolymer gels, *Food Hydrocolloids*, 11, 159, 1997.
75. Tolstoguzov, V. B., Thermodynamic aspects of dough formation and functionality, *Food Hydrocolloids*, 11, 181, 1997.

76. Sahi, S. S., Interfacial properties of the aqueous phases of wheat flour doughs, *J. Cereal Sci.,* 20, 119, 1994.
77. Kalichevsky, M. T. and Ring, S. G., Incompatibility of amylose and amylopectin in acqueous solution, *Carbohydr. Res.*, 162, 323, 1987.
78. MacRitchie, F., Baking quality of wheat flours, *Adv. Food Res.*, 29, 201, 1984.
79. Eliasson, A.-C. and Larsson, K., in *Cereals in Breadmaking*, Marcel Dekker, New York, 1993, chap. 2.
80. Ring, G., Colonna, P., l'Anson, J., Kalichevski, M. T., Miles, M. J., Morris, V. J., and Orford, P. D., The gelation and crystallization of amylopectin, *Carbohydr. Res.*, 162, 277, 1987.
81. Muller, H. G., Application of the statistical theory of rubber elasticity to gluten and dough, *Cereal Chem.*, 46, 443, 1969.
82. Bushuk, W. and Hlynka, I., Water as a constituent of flour, dough and bread, *Baker's Dig.*, 38, 43, 1964.
83. Zobel, H. F., A review of bread staling, *Baker's Dig.*, 47, 52, 1973.
84. Knyaginichev, M. I., The staling of bread and the maintenance of its freshness, *Starch,* 22, 435, 1970.
85. Willhoft, E. M. A., Bread-staling I and II, *J. Sci. Food Agric.*, 22, 176 and 180, 1971.
86. Piazza, L. and Masi, P., Moisture redistribution throughout the bread loaf during staling and its effect on mechanical properties, *Cereal Chem.*, 72, 320, 1995.

2 Plasticization: The Softening of Materials

Attilio Cesàro and Fabiana Sussich

CONTENTS

0-8493-8790-6/01/$0.00+$.50

Natural things live in a dynamic instability;

"the order parameters are ultimately the thoughts."

Haken[1]

BREAD: A POLYMER POINT OF VIEW

It is commonly asserted that a dough product is a continuous network of an amorphous polymer (gluten) filled with partially crystalline starch granules. The weight of the filler is much higher than that of the network-forming component. At some stage of the preparation, a bakery product can be considered from the macromolecular point of view as an *interpenetrating polymer network* (IPN). This definition implies that the system is composed of a combination of two or more polymers in a network form, where at least one of the polymers is polymerized and/or cross-linked in the immediate presence of the other components.[335] It is also recognized that the fundamental aspect of the IPN is the immiscibility of the polymers and the system must be phase-separated. The stage and conditions (kinetics, history, initial composition, and temperature) for the occurrence of phase separation determine the morphology of the final product. Once the structure has been formed, the mobility of low molecular weight components and the local mobility of high molecular weight polymers continue to modify the properties of the system with time. As we shall see in this chapter, the concepts behind plasticization are the best way to rationalize the behavior of these changing properties. Different molecular species (those with different chemical structures) exhibit chemical and physical properties which result directly from the interaction between all the constituent atoms of the molecule. However, it has become clear in recent years that a fundamental and often forgotten aspect of most practical applications is molecular morphology; that is, the way the various molecules are topologically organized as a result of the sum of many weak interactions. The aim of this chapter is to guide the reader through the changes in the properties of the dough system, which are the consequences of compositional rearrangements and physical variables within the framework of polymer morphology at the molecular level.

As a starting point, let us consider the commonly accepted chemical composition of the dough system (Table 1). We must first distinguish between low and high molecular weight components. This is fundamental to the polymer approach for understanding the property changes in the system. The solid structural matrix of bread is composed mainly of polymers, which can be either completely amorphous (gluten) or partially crystalline (starch), the latter especially when the bread is aged after baking. Microstructural approaches reveal that the state of the system is more similar to a composite than a homogenous material.

TABLE 1
Typical Compositional Data for Bread Dough

a) Davidou et al., 1996[3]

High Molecular Weight		*Low Molecular Weight*	
Wheat flour	50.5	Water	41.5
Bakers' yeast	2.4	Salt	1.1
Pastrycook butter	2.4	Sugar	1.6
		Calcium propionate	0.5

b) Piazza et al., 1995[4]

	Yeast Bread	*Sourdough Bread*	*Lactic Acid Bread*
Flour (g)	100.00	100.0	100.00
Water (ml)	60	48	60
Yeast (g)	3.75	4.3	3.75
Salt (g)	1	1	1

c) Tolstoguzov, 1997[5]

Proteins	11%
Soluble pentosans	0.8–1.2%
Damaged starch	3%
Water-soluble proteins	2% (albumins and globulins)

Notes:
a) – g/100g wet basis for control white bread.
b) – For different bread the dough composition changes slightly.
c) – Beside the native starch granules, wheat bread flour contains the listed polymers.

DEFINITION OF TERMS

This chapter focuses on how a material can be modified by the presence or addition of another component. The modification does not implicitly require the presence of a second component in addition to the basic plastic material. The terms plasticity, plastic deformation, and plastic are typically derived from the viscoelastic (mechanical) world, and must be clearly defined for end-use properties. They are all used to transmit a sense of deformability, i.e., a time-dependent property. From the science world, these terms have pervaded the visual perception of artistic expression and the social sciences. The picture of Mirone's famous "Discobolos"[6] elegantly shows the tension instability of the player which becomes effectively plastic when seen from a different position rotated at about 90° (Figure 1). Other excellent examples of visual plasticity are Escher's geometric compositions. However, Figure 2 is probably the best example of the artist's intention to represent the dynamic interaction between the instrumental method, "the eye", and the material under observation. In this case, the persistence of a tri-dimensional image is only related to the dynamic control of the sensorial stimulus of the observer.[7] In both cases, however, plasticity is a visual perception of a changing world. On a scientific basis, the definitions should only refer to quantitative changes of physical properties of a polymeric material.[8]

FIGURE 1 The perception of changes as shown by Mirone's Discobolos is called plasticity (Museo Vaticano). Reproduced with permission of Garzanti Editore.

Plasticity

Plasticity is the capacity of a material to be molded and to retain its shape for a significant period of time under finite forces. Materials exhibiting plastic behavior are described as soft solids, as they exhibit flow above a yield stress, passing from the elastic to the plastic deformation. In the most general sense, a "plastic material is capable of being shaped, through plastic flow, by the application of deforming forces."[8] In a polymer context, such material is "based on a high molecular weight polymer and may be distinguished from a rubber by its high stiffness and lack of a large reversible elastic deformation, although no sharp division can be made between plastics and rubbers."[8] The distinction between plastics, fibers, and coatings rests on the physical shape of the product.

Plastic Deformation

Plastic deformation occurs after yielding. In the original definition, as applied to metals, the deformation is considered irreversible, as opposed to the "completely

FIGURE 2 *Plastic Permutation* by F. Grignani (1959) has been interpreted as the perception of dynamic instability. The relaxation of sensorial stimulus determines the plastic permutation. Reproduced with permission of Mrs. Jeanne Grignani.

reversible deformation that occurs elastically before yielding. It may thus be considered truly as plastic flow."[8] However, for polymers, post-yield deformation can be wholly or partially recoverable. Nevertheless, "in the glassy state, polymers may approximate to ideal plastic behavior, so the concepts of classical plasticity may sometimes be usefully applied."[8]

Plasticizer

Plasticizer is a substance (diluent) that is incorporated into plastic materials to improve their workability and increase flexibility. Polymer and plasticizers should form a homogeneous single-phase mixture; mixing of the two components is necessary to lead to substantial physical property modifications with changing plasticizing content. The plasticizer is usually a liquid, but occasionally is a low melting point or softening point solid, which solvates a polymer and therefore softens it, i.e., acts as a flexibilizer. To be useful, a plasticizer must also exhibit permanence so it will not be lost during use. Therefore, practical plasticizers are high boiling-point (sometimes high molecular weight) organic liquids which are of a similar solubility to the polymer and are therefore compatible with the polymer being plasticized. A plasticizer also lowers the T_g value and the softening point of the polymer and thus allows for easier processing. When small quantities are used specifically for this purpose, as is common for processing rubbers, plasticizers are referred to as process aids. In a broader definition, plasticizers are classified as primary (compatible over the whole composition range), secondary (of limited compatibility), or extenders (only compatible when used in combination with a primary plasticizer).

Cross-linkers and Fillers

There are other additives which modify the mechanical features of a polymeric material. On heating, all linear polymers flow. Their viscosities decrease steadily to reach the melt points. That is why they are termed thermoplastic. The simplest way to avoid this viscosity change with temperature is to increase the molecular connectivity in a network by cross-linkage. One of the best known cross-link reactions is rubber vulcanization. The other, less well-known in the synthetic polymers field, is the cross-link reaction in gluten. Cross-linkers are reactive molecules which can be added to the system to bind the chains together to form a network. The product is called a thermoset, because it does not flow on heating.

Relaxations

The word relaxation characterizes the transformation processes (usually slow) which occur in a system moved away from the equilibrium. While reversibility of a transition is the principle which identifies its thermodynamic character, hysteresis phenomena show that potential barriers are present, causing a damping effect on the transformation. In many cases, the rate of molecular transformation could be so slow that understanding the thermodynamic stability of the system becomes complex. Slow transformations may occur at constant temperature in the absence or presence of external forces (mechanical or electrical). If the applied force is modulated with a periodic oscillation, the properties measured may change within a particular range of the frequency of modulation as a relaxation response to the modulated force. Relaxations (frequencies) and transitions (temperatures) are the macroscopic manifestations of the same molecular phenomena, that is, the presence of molecular motions in the system. The importance of molecular motions on the property of a material lies in the performance of the material; for example, mechanical stresses, which result in reversible deformations and no fractures, can be tolerated if the system can dissipate these forces at the molecular level. Additional information can be obtained by knowledge of the transitions temperature range (thermal analysis, DSC, dilatometry) and frequencies (DMA, DETA, NMR) characteristic of a material (see Chapter 5).

PHASE TRANSITIONS OF POLYMERS. I: MELTING AND CRYSTALLIZATION

General Principles of Phase Transition in Polymers

In addition to the equilibrium thermodynamic states classically studied (i.e., crystals, liquids, and gases), there are other states (mesomorphic or with different morphologies) important in practice. Polymers and other compounds present a number of molecular features, since they can have non-regular chemical structure and cannot crystallize and, therefore, do not melt. However, non-regular polymers do present other phase transitions. These transitions invariably change the sensorial characteristics of the polymer and can be revealed by many instrumental methods, although they require careful preparation and well-designed experimental measurements.

Virtually all polymers present a glass-rubber transition phase (glass transition) wherein the disordered polymer softens above the temperature of transition T_g. This transition is associated microscopically with the onset of long range, cooperative molecular motions. In addition to the occurrence of a melting and a glass transition, there are other less distinct changes in some mechanical, thermal, (di)electrical, and molecular properties which are consequences of a change in the states of molecular association and of their dynamics. Since the effect of plasticizers is general and similarly affects all these transitions, these states will be reviewed, focusing particular attention on some of the widely accepted phenomena. The explanation of the fundamental aspects of polymer transitions and the intrinsic factors which affect them is given to reveal a unifying approach to the problems with the plasticization of the polymeric systems.

The Melting of a Polymer

While amorphous polymers do not contain crystalline regions by definition, crystalline polymers are generally only about 30 to 60% crystalline and contain appreciable amounts of amorphous material, the extent of which depends mainly on the thermal history of the sample (i.e., how the solid was prepared). When a polymer is heated above its melting point, the crystalline part melts into an amorphous liquid. Its transformation back to the solid phase upon cooling and its reorganization into a partially crystalline material are matters of thermodynamics and the kinetics of crystallization in an often very viscous phase. In the absence of non-equilibrium phenomena, the melting and crystallization processes generally occur as a first-order transition at the equilibrium temperature, T_m°, defined by the ratio of the enthalpy and entropy changes at the transition, $T_m^\circ = \Delta H_m^\circ / \Delta S_m^\circ$. A first-order transition is defined on a thermodynamic basis by equality of the molar Gibbs free-energy G of the two phases at the temperature T_m° (Figure 3a), whereas the first derivatives of G (with respect to the temperature or pressure) for the stable phase show a discontinuity step at T_m° (Figure 3b). This thermodynamic equilibrium temperature refers to a perfect crystal of very large dimensions with no surface effects. However, the presence of defects or a decrease in crystal size invariably produces a lowering of the melting temperature. All pure polycrystalline materials (i.e., crystalline solids in which the dimensions of the crystals are very small) invariably present a lower melting point with respect to common crystalline systems with crystals of visible dimensions. Pre-melting of polycrystalline ice in grain junctions (impurity-free) has been experimentally determined and theoretically predicted in literature.[11] The situation for polymers has been known for a long time. Reproducible data illustrates how crystal size and melting temperature depend on the experimentally-determined value of the lamellae thickness. The change in size is therefore limited to a one-dimensional variable. The Hoffman and Weeks equation correlates the melting temperature with the crystallization temperature,[10] recognizing that crystal growth depends on the temperature to which the liquid is cooled below T_m° (undercooling).

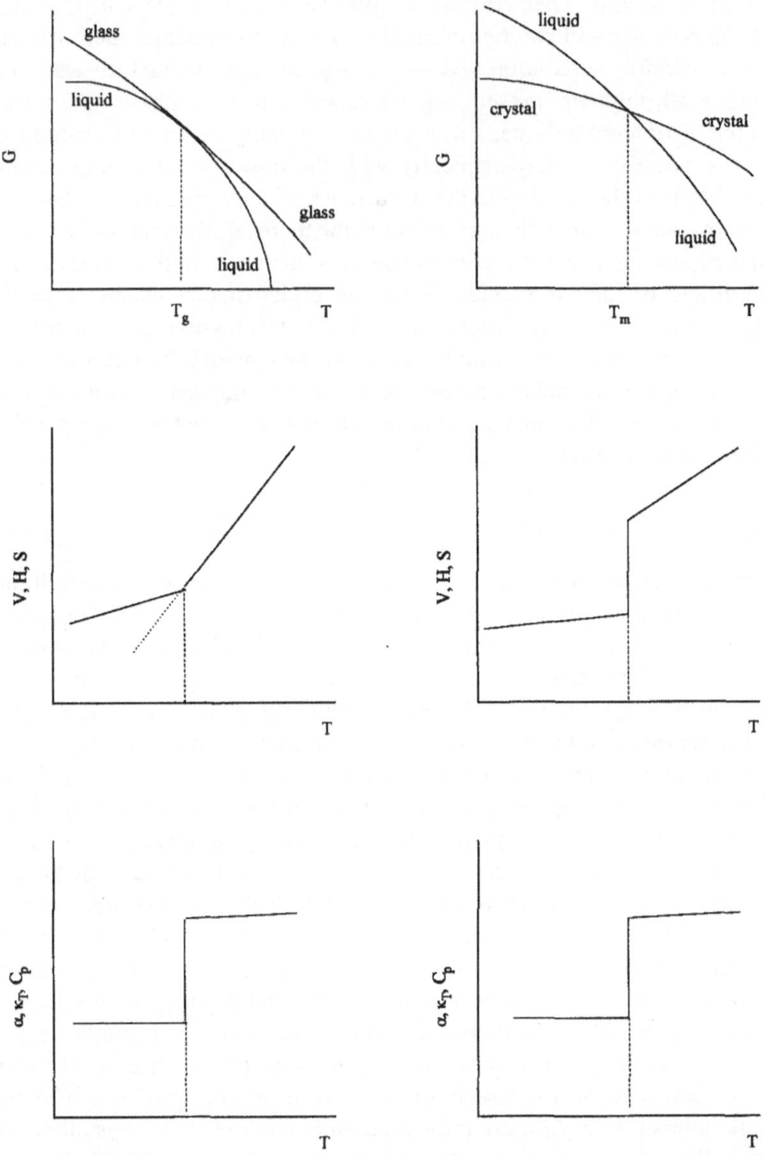

FIGURE 3 Schematic temperature dependence of the thermodynamic functions for a first and a second order transition. a) Gibbs free-energy (G), b) first derivatives (volume, V, enthalpy, H, and entropy, S), and c) second derivatives (expansion coefficient, a, isothermal compressibility, k_T, and heat capacity, C_p).

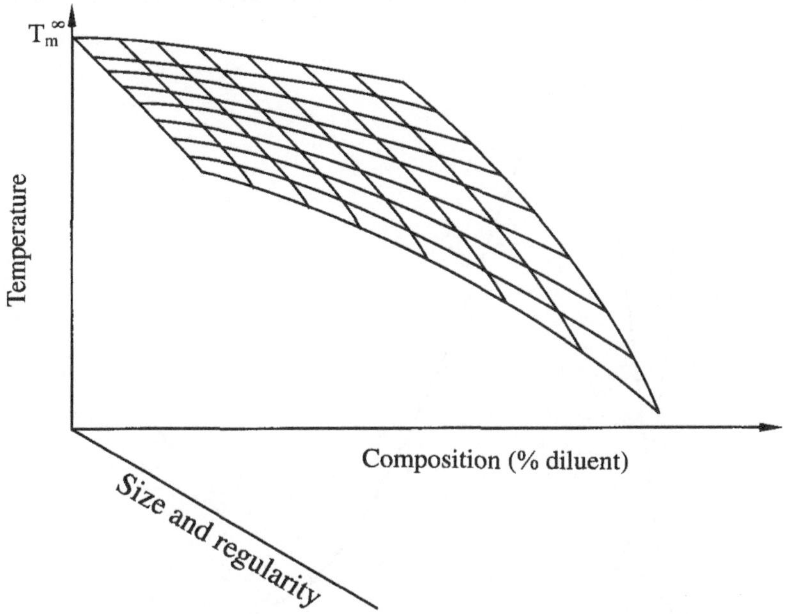

FIGURE 4 Schematic tridimensional diagram of dependence of the melting temperature of polymer crystals as a function of composition of the liquid phase and decreasing size and regularity of crystals (according to Gibbs free-energy changes calculated by equations of Flory-Manderkern and Hoffmann-Weeks).

The melting point depression, which is commonly observed for low molecular-weight crystalline substances in the presence of another component, contributes to and amplifies the other effects due to the presence of irregular and small polymer crystals. A schematic tridimensional dependence of the changes of Gibbs free-energy and of the melting temperature as a function of the perturbations considered here is given in Figure 4.[11]

In summary, if a polymer presents diffuse irregularities in its chemical structure (substitutions, branching, or configurational defects) the crystallizable fraction may be very low, the regularity of the crystals may be affected, and a large depression of the melting temperature could occur. The conditions which lead to nucleation of crystals but hamper the steady growth of the nuclei (e.g., large undercooling temperatures which increase the viscosity and hamper the diffusion processes necessary for the crystal growth) produce a material made of thin crystals. A large depression in the melting temperature of the final product is then detected. The ideal melting temperature can also be depressed by the addition of a component (or impurity) which is miscible with the polymer liquid phase. Low molecular weight molecules, monomers, and plasticizers all belong to this category if they are completely compatible with the liquid-amorphous polymer phase. This description relies on a knowledge of the crystallization process, its kinetics, and the morphology of the crystalline phase which is produced.[11,12]

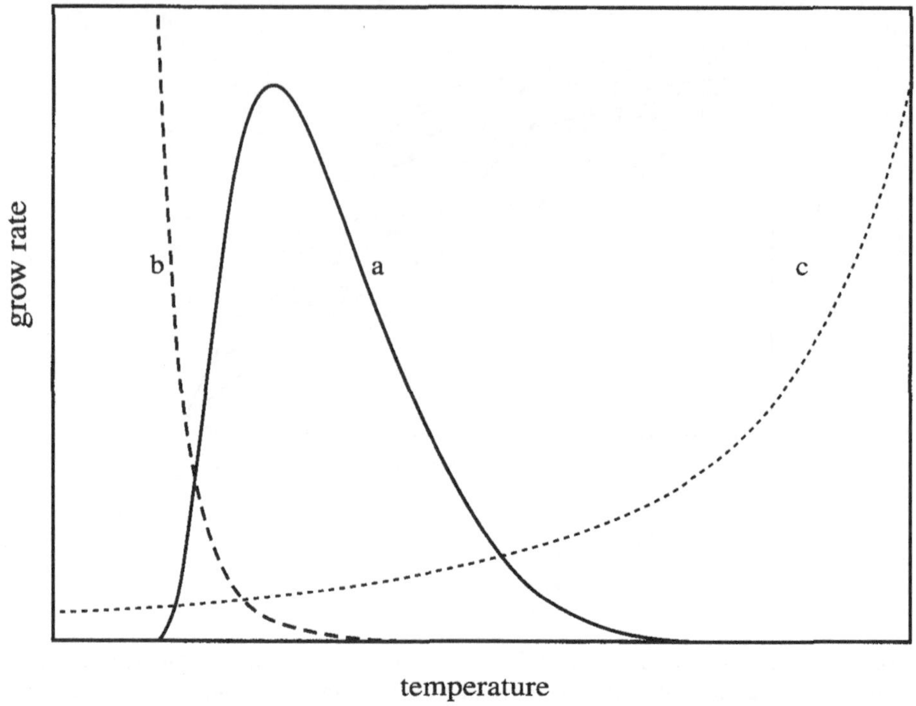

FIGURE 5 Plot of the linear growth rate (a) versus crystallization temperature. For this polymer, T_g = 275°C, and T_m = 350°C, at which points the rates of crystallization are zero. The theoretical trends of nucleation rate (b) and diffusional rate (c) are also shown.

Crystallization Processes

Besides the model of crystallization, a brief account of the kinetics of crystalline formation is relevant. Ultimately, the properties of the crystalline fractions we are studying are a consequence of the conditions under which the crystals are formed. Of these variables, the thermodynamic stability of the crystals and the mobility (diffusion) of the segments that form the crystals are both functions of temperature.

Undercooling is always necessary to induce crystallization. This is explained on the basis of the thermodynamic stability needed for the nuclei. A schematic dependence of the formation rate of stable primary nuclei on temperature shows that it reaches a maximum value below the melting temperature (Figure 5). The sharpness of the bell-shaped curve depends on the material (the potential barrier to formation of stable nuclei and the potential barrier to segmental diffusion). The two zero-limiting values at low and high temperatures are easily understood. Overall crystal growth ability in constrained domains decreases as the size of the domains decrease,[13] based on statistical probability of nucleation events. It has been suggested that this phenomenon occurs in the native microbial granules of poly(hydroxybutyrate) and could explain the unusual *in vivo* amorphous state of the pure polymer, despite its tendency to undergo crystallization when extracted from cells.[14]

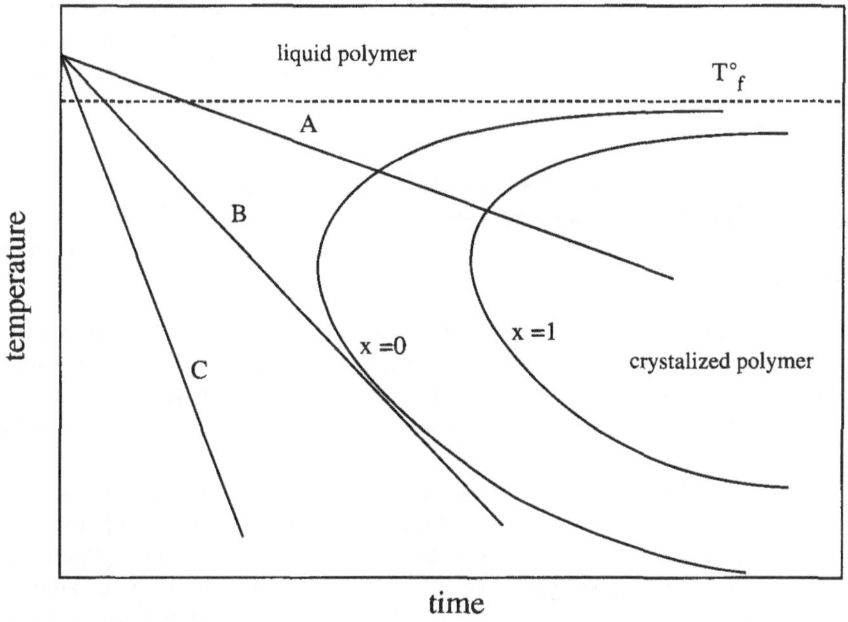

FIGURE 6 The diagram temperature-time shows the relationship between cooling rate and the phase reached. Three regions can be defined: liquid polymer at high temeratures; undercooled liquid phase at high cooling rates (dT/dt > B); and crystallized polymeric phase. The region between curve x = 0 and x = 1 contains an increasing amount of crystalline fraction, x. Cooling rates increase from A to C.

A practical way to display the kinetic and thermodynamic dependence of crystallization processes is a diagram which shows the extent of crystalline formation plotted against temperature and time; note that a line in this diagram (T versus t) indicates a process with a scanning rate dT/dt. Figure 6 shows three schematic cooling processes, A, B, and C. Only A, the slowest, produces complete crystallization. Complete crystallization does not refer to 100% of the material but only to that part of the material which can be crystallized under appropriate conditions. A specific case of crystallization may be obtained by forcing the chain molecules to crystallize in an oriented fashion under elongational strain. The kinetics of crystallization of a melt under elongational flow is faster and oriented fibrillar crystals present a higher melting temperature than unoriented crystals.[15]

The melting of a crystalline polymer corresponds by definition to loss of order, which can easily be detected by diffraction methods (x-ray, electron, or neutron). The sharp lines or spots characteristic of the crystalline layers change to diffuse halos of the liquid or amorphous state. Other direct investigations include optical polarized microscopy, which shows the textural changes of the crystalline phase to a nearly structureless field. All other common methods are thermodynamically based for determining properties (i.e., volume and enthalpy) which are functions of the state of the system. Many of these approaches are not sufficient to define the type of transition, but their sensitivity and simplicity of operation make these methods most popular. These quasi-static methods illustrate the presence or transformation of that portion of the polymeric system which is organized in a crystalline fashion. This does not mean that we have to ignore the non-crystalline portion, since time-dependent modifications are easily ascribed to the amorphous fraction.

PHASE TRANSITIONS OF POLYMERS II: GLASS AND SECONDARY TRANSITIONS

Glass Transition

The glassy state is a conformationally disordered polymer system in which the cooperative chain motions are frozen, so that only limited local motions (such as side-group rotations) are possible. The glass transition temperature is the temperature at which the transition occurs from this state to a state in which molecular motions become accessible.

The non-equilibrium flavor of the glassy state has been scientifically established. The transformation to a glass occurs on cooling a liquid below a temperature T_g if the molecular rearrangements necessary to accommodate the decreasing temperature slow down sufficiently, because these motions would require a longer amount of time than the imposed cooling rate allows. This definition implies a frozen-in structure, which may be characterized below T_g and will be retained as long as the same cooling rate continues. The properties may differ substantially from the outside to the inside of any massive piece of amorphous/glass material cooled faster than 10^{-3} K min^{-1}. In all elastic-modulus and relaxation experiments the rate and amplitude of deformation, the thermal history of the sample, and the instantaneous temperature must be defined. This method defines T_g as a discontinuity in a multidimensional coordinate system.

Summary of Glass Transition Theories

The basic impact of glass transition theories can be summarized by reporting a few of the hypotheses. The theory of free volume (first developed by Eyring[17] and applied specifically to liquid viscosity by Doolittle[18]) introduces this concept to describe the viscosity of liquids as an activation energy for the diffusional rate. Although there is no rule for calculating the free volume, these cavities in the form of segment-size voids are the requirements for the onset of coordinated molecular motion. The theory provides relationships between coefficients of expansion below and above T_g and yields equations relating viscoelastic motion to the variables of time and temperature.

The Doolittle equation provides the basis for the Williams-Landel and Ferry (WLF) equation,[19] which is the analytical relationship between polymer melt viscosity and free volume. Its strength lies in its generality: no specific chemical structure is assumed other than a linear amorphous polymer above T_g. The kinetic flavor of the WLF theory is introduced into the formulation of the kinetic theory of glass transition. The kinetic theory of glass transition considers molecular and macroscopic response within a varying time frame. The material is said to be in the glassy state if the number of holes and their spatial positions become frozen in; that is, the molecules are unable to move from their location into a hole. The kinetic theory defines T_g as the temperature at which relaxation time for segmental motions in the main polymeric chain is of the same order of magnitude as the time scale of the experiment. Therefore, the kinetic theory is concerned with the rate of the system's approach to equilibrium, taking the respective motions of the holes and molecules into account. The kinetic theory also provides quantitative information about heat capacity below and above the glass transition temperature, and explains the 6 to 7°C shifts in glass transition as the experimental time scale increases by a factor of ten.

The thermodynamic theories introduce the notion of equilibrium and the requirements for a true second-order transition, albeit at infinitely long time scales. The theory postulates the existence of a true second-order transition, which the glass transition approaches as a limit when measurements are carried out more and more slowly. It successfully predicts the variation of T_g with molecular weight, cross-link density, diluent content, and other variables. In the hypothetical infinite time scale experiment, Gibbs and DiMarzio[20] provided a thermodynamic answer to the question of the equilibrium properties that a glassy material should have. The theory is based on a lattice model, and the underlying true transitions can possess equilibrium properties, even if they are difficult to realize.

Although there is no lack of theories of vitrification, none of the proposed conceptual models have successfully accounted for the observed phenomena.[21]

The following empirical rules can be safely stated[11,22] which combine practice with theory:

- The experimental value of T_g appears at progressively lower temperatures as the experiment is conducted at a slower rate
- Sufficient time must be allowed for reaching thermodynamic equilibrium in many simple reactions
- Changes in the conformational state of the polymer chain backbone occur much more slowly (near the T_g) than most other molecular processes which are often given as examples of simple equilibria
- If a true equilibrium T_g value exists, it must be a lower value than observed in short-term experiments

Thermodynamics is still fully applicable within the dynamic definition of the glassy state. It must, however, be remembered that for many simple chemical reactions, sufficient time must be allowed for thermodynamic equilibrium to be reached. The instability of the glassy state dictates that if cooling is arrested, the material will have excess thermodynamic properties (chemical potential) which are the driving force for approaching equilibrium. The phenomenon of this slow process is called physical aging and occurs around 15°C below T_g, since the time scales for reaching equilibrium increase rapidly with the decrease in temperature (see the temperature dependence of diffusion process). A concise description of the fundamental problems related to phase transition and glassy state is found in Bondi and Tobolski.[21] A more detailed review of the slow modifications occurring in the glassy state is given by Hutchinson.[16]

Independent of the theoretical formulation of glass transition at molecular level, these generalizations can be made about the correlation between the molecular structure of the system and the experimentally determined value of T_g. Experiments must be carried out under suitably controlled conditions, with comparisons between reliable data. The most important generalization from the polymer point of view concerns the local dynamics of the chain (often called flexibility). However, even in consideration of the ergodic theorem in which static random conformation of polymers implies dynamic changes with time, transformation time also remains an important factor.

The following categories of structure-property correlations[12,22] are provided with the aim of rationalizing the experimental findings on the basis of local structural modifications and macromolecular parameters.

1. A low T_g value is expected for a freely rotating polymer chain with weakly interacting forces and without bulky side groups (i.e., poly(dimethylsiloxane) with a T_g of –123°C).
2. The insertion of more rigid co-monomers along a flexible chain raises the T_g. For example, the T_g of poly(ethylene) changes from a range of –80°C to –125°C to +70°C for poly(ethyleneterephthalate), but there is some debate in the literature about the true value of T_g.[22] A similar effect is produced when side groups are regularly present in a series of homologue polymers as it is clear that, in going from poly(ethylene) to poly(propylene) and to poly(styrene), the T_g increases to –18°C and +100°C. The consequence of the presence of side groups is a decrease in local conformational freedom and therefore better intimate contact between different parts of the chain.
3. The presence of flexible side groups offers an interesting example of improving cooperative molecular motions which cannot otherwise exist for an inherently rigid chain such as a soft cover on a rigid body. This is the case of the series of poly(n-acrylate)s and poly(n-methacrylate)s, with flexible n-alkane pendant groups. Similarly, in the family of polysaccharides, the best way of reducing the T_g is to introduce linear hydrophobic ether or ester derivatives as side chains.[23] In this case, the internal plasticization derives from the flexibility of the side chains and the reduction of the intermolecular (polar) interactions which are responsible for the cohesive energy of polar systems (and, as a consequence, high T_g).
4. The resulting effect of more substituents on the same main backbone is complicated. If the substituents are small, the prevailing result is an increase in chain rigidity and therefore an increase in T_g (see 2). However, regular substituents may change the intrinsic stiffness of the chain in the opposite direction, as is the case of cellulose tricarbanilate,[24] which shows a much higher flexibility than the unsubstituted polymer.
5. Staying within the same class of polymers (i.e., the same monomeric species), the first relevant parameter which affects T_g is the molecular

weight (MW), since the local dynamics of chain ends are always much greater than those of the core monomers.[25] An empirical equation relates T_g to M (as well as for the melting temperature):

$$T_g = T_g^{\infty} - K/M \tag{1}$$

where T_g^{∞} is the transition temperature for a polymer with an infinite M, and K is a constant.[26] This relationship holds for linear oligomers and polymers and has also been illustrated in the case of linear amyloses and oligodextrins.[27-29]

6. Non-linear chains (branched, star, combs, etc.) show a reduction in T_g as a consequence of increasing the number of chain ends with respect to the related linear polymer that has the same M. Among the polymers of interest to us, this is demonstrated by the structurally different linear amylose chains and branched amylopectin tree-like chains.[29] Conversely, cross-linking agents (either chemical or physical) increase the connectivity between different chains and decrease the number of loose ends. T_g increases in line with the true (chemical cross-links) or apparent (physical cross-links) molecular weight.
7. The dependence of the T_g of co-polymers varies according to the type of microstructure of the chain. Alternate co-polymers fall into category 2. Block co-polymers show two different values of T_g, each corresponding to the value of the corresponding homopolymer. Block co-polymers usually behave as an intimate mixture of two species. Random co-polymers show a T_g that appears to be a suitable average of the two T_gs of the corresponding homopolymers. This is dependent on the co-monomer composition, and changes from one T_g to the other. An acceptable approximate equation, still widely used, is that originally formulated by Fox[31]:

$$\frac{1}{T_g} = \frac{w_1}{T_{g1}} + \frac{w_2}{T_{g2}} \tag{2}$$

where w_1 and w_2 are the mass fraction of the two co-monomers.

Although not strictly structure–property related, the presence of low molecular weight molecules dilutes the polymer chains and increases molecular mobility. The term plasticization is invariably associated with this effect. However, in some cases when low molecular weight substances are added to a polymer, the T_g of the polymer may also increase (anti-plasticization). This apparently contradictory result must be accepted based on point 6, where a physical cross-link may occur as a result of the addition of low molecular weight solutes (for example, calcium ions added to a polycarboxylate). The presence of a diluent is the last condition for the scope of this chapter.

Some examples modifications occurring in the T_g values, as illustrated in the previous schematic list, are shown in Figures 7 and 8 and in Table 2.

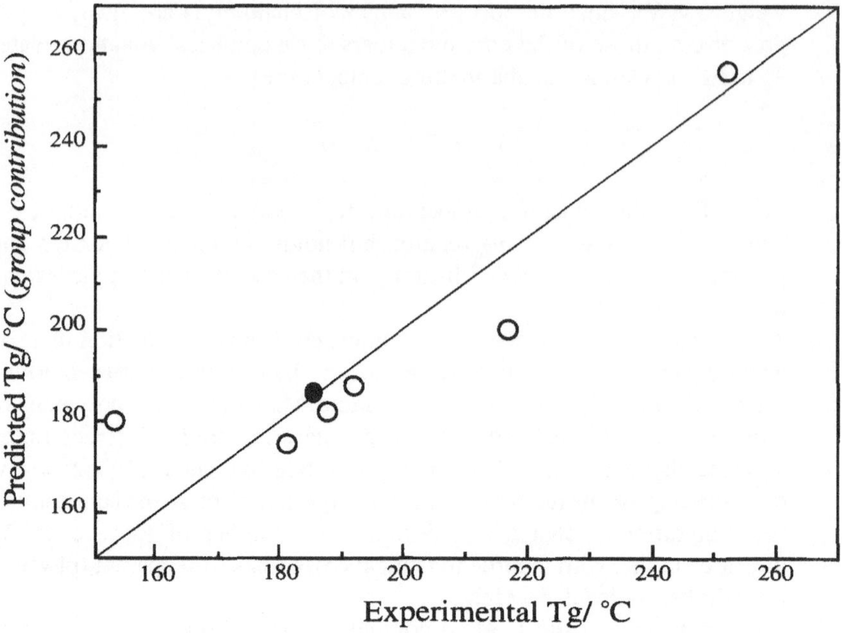

FIGURE 7 Predicted glass transition temperature for food proteins based on additive contribution method.[62] The full circle is the estimated T_g for gliadin (186°C).

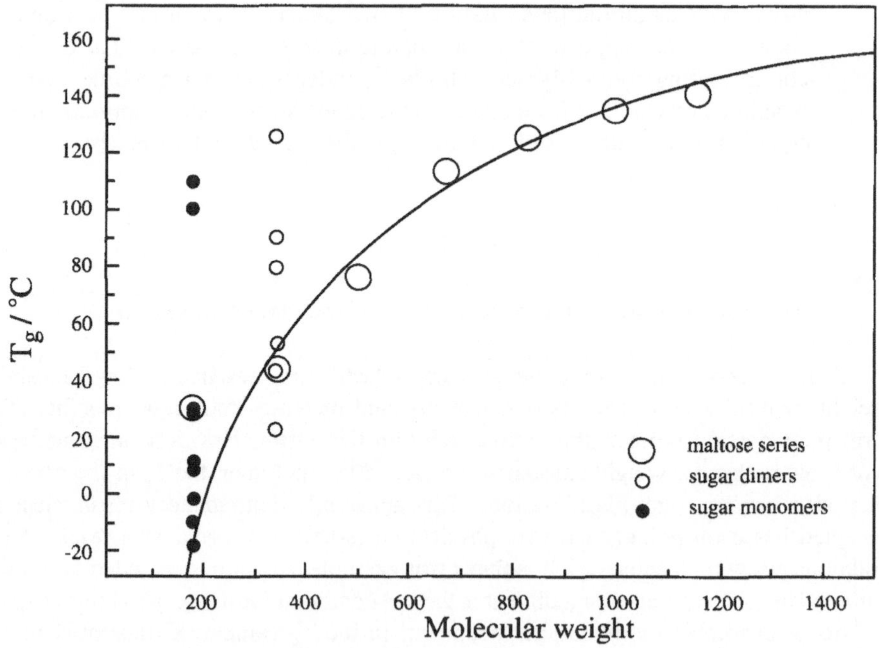

FIGURE 8 Molecular weight dependence of the glass transition temperature for low molecular weight sugars and for maltose series.

TABLE 2
Selected Values of T_g (°C, mid-point, DSC) for Mono and Disaccharides

Monomers		Dimers	
Galactose	36	Sucrose	68.5
Fructose	10	Maltose	92
Glucose	37	Cellobiose	77
Mannose	31	Trehalose	125
Altrose	10.5	Mannobiose	90
Sorbitol	–4		
Xylitol	–23		
Xylose	13		
Arabinose	3		
Ribose	–10		

Molecular Relaxations below the Glass Transition Temperature

Although the glassy state is characterized by absence of large scale molecular motions, local motions (secondary transitions) still appear possible below the glass transition temperature. These motions involve only partial movements (oscillations or partial rotations) of small segments of the polymer chain. Since they do not change the energy or volume of the system significantly, they cannot be detected by usual calorimetric or dilatometric measurements. Other physical properties must be studied, mainly by techniques based on periodic oscillations. For example, typical viscoelastic behavior of polymeric materials can be quantified by determination of the elastic (storage) component and the viscous (loss) component. Measurements of these quantities as a function of temperature (in the appropriate frequency range) invariably show relaxation peaks with a characteristic transition temperature, although the whole transition can be spread over 50 to 100°C. More than one relaxation process is usually detected in addition to the glass transition process by using, for example, dynamic-mechanical methods. These transitions are indicated by Greek letters. The symbol α is assigned to the glass transition and the symbols β and γ to the transitions that occur as the temperature decreases.

Experimental data show that most secondary relaxations in glassy polymers are a consequence of conformational isomerization of short sections of main chains or side chains. Their kinetics may be described by the site model in which stable conformations are separated by a potential barrier (Arrhenius behavior). Secondary relaxations can be studied using dynamic methods based on the interference of an oscillating external force field with molecular motion. However, because of this kinetic character, data on the location of these relaxations on the temperature scale must be supplemented by frequency of measurement when it cannot be referred to a common frequency value, e.g., 1 Hz. The current models do not give a satisfactory explanation of the mechanism of energy losses at a molecular level.

Molecular motions causing secondary relaxations in glassy polymers were first segregated by Heijboer into the following types[32]:

1. Local main-chain motions assuming rotations of up to six groups around co-linear bonds at the two ends of the segment
2. Side-chain rotations around the bond linking the group to the main chain
3. Internal motions within the side chain
4. Molecular motions of, or affected by, a diluent.

Although the last group is the least studied so far, it requires further specification for the scope of this chapter. Since we are discussing secondary relaxations, the polymer phase must be glassy by definition, i.e., the system is below the T_g.

Diluent-induced relaxations[33] are assigned to the motions of complex units which consist of a group of the polymer segment associated with molecules of the diluent. They can be distinguished according to the type of motion exhibited by the same polymeric group in the absence of the diluent. However, we shall see that, although not exclusive to food polymers, mobility changes of the secondary transitions measured through the temperature of the relaxations can be very large and frustrate all attempts of correlation.

DEPENDENCE OF A TRANSITION TEMPERATURE WITH FREQUENCY

As stated in the third law of thermodynamics, all crystallizable materials prefer ordered states at low temperatures. This implies that all properties of the system are in internal equilibrium. This is far from true in terms of practical results, as illustrated by the occurrence of the glassy state and the incomplete crystallization of polymers, even if supported by their structural regularity. However, the properties of a non-equilibrium material may often appear to be in a stationary state if the instrument measuring time is short with respect to the slow transformation which characterizes the instability. The conventional classification between the equilibrium state and the non-equilibrium state must be replaced with a distinction between equilibrium-like and non-equilibrium-like properties of the material. This distinction can be achieved by using an operational quantity for the measurement of time-dependent phenomena, the "Deborah number", De, introduced by Reiner,[34] which is defined as the ratio between molecular relaxation time of material transformation and the characteristic observation time of the instrument. The material can only be classified, although still ambiguously, with a given characteristic time in an equilibrium or non-equilibrium state. If a large range of time is considered, the entire curve of the changing property associated with molecular transformation toward the equilibrium should be observed, provided that the interval from $De \ll 1$ to $De \gg 1$ can be explored. For a given transformation process of material with respect to $De \ll 1$, all molecular motions are faster than the observation time and the material appears to be in a dynamic equilibrium that is conventionally said to conform to the ergodic state. As temperature decreases and relaxation times of the material increase, the condition $De = 1$ will be encountered. A glassy state would, therefore, be characterized by a value of $De \gg 1$.

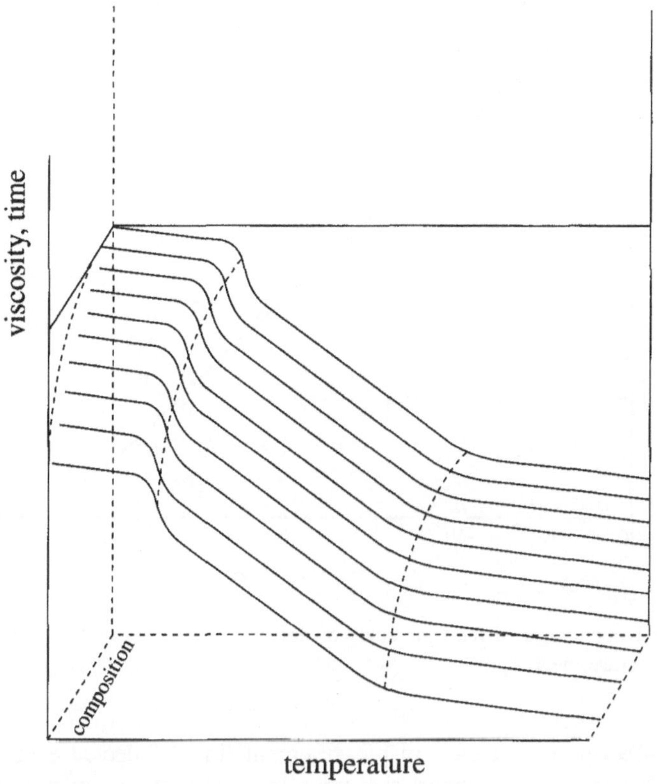

FIGURE 9 Schematic tridimensional plot of the dependence of viscosity on temperature and composition. The effect on the T_g is illustrated.

A characteristic time scale of the viscous flow of the material can be defined empirically to distinguish amorphous fluid-like material from glassy solid-like material, even in the presence of a diluent (Figure 9). In this way, the calculation of the temperature at which the amorphous/undercooled liquid exhibits a viscosity of $\eta = 10^{13}$ (P·s) yields a conventional T_g. This conventional time scale defines the T_g by means of an iso-viscous state.

This section is dedicated to the relaxation processes manifested when the frequency of mechanical manifestation becomes comparable to the frequency of the relaxation characteristic of the molecular motion under consideration. We have shown that transition phenomena related to slow motions are affected by the instrument operation time. To illustrate this, Figure 10 shows the dependence of the T_g of polystyrene[22] on the observation time. We return to this concept to examine some of the consequences in greater depth and to illustrate the necessity of this non-ancillary information. This leads to the conclusion that the data on non-equilibrium transition may be of little use if it lacks the specification of the appropriate range of frequencies.

For viscoelastic properties, this correlation between time and temperature is called time-temperature equivalence and implies that an increase in temperature produces

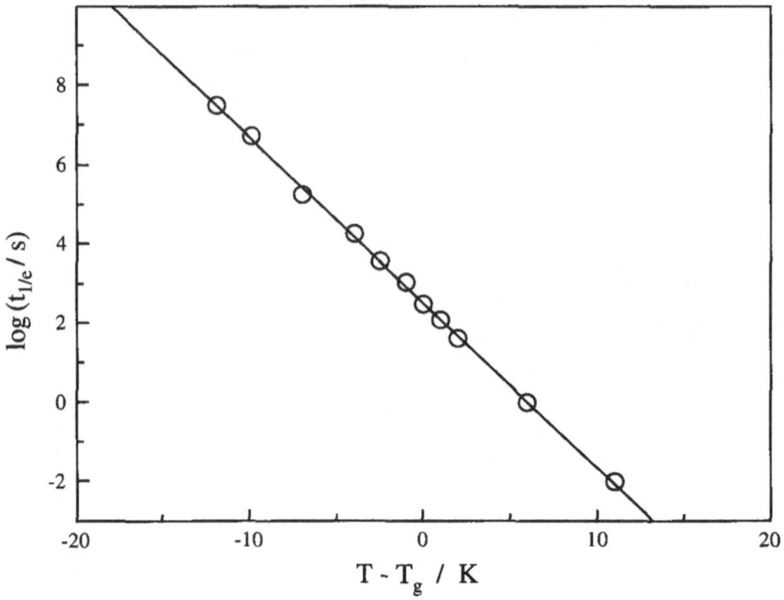

FIGURE 10 Dependence of the time scale of deformation (measured as $t_{1/e}$) of polystyrene on the temperature, near T_g.[36]

the same effect as an increase in measurement time or decrease in measurement frequency. It can be easily shown that upon increasing the frequency ($n_1 < n_2 <$ n3) of measurement, the entire viscoelastic relaxation moves to a higher temperature ($T_1 < T_2 < T_3$). This dependence is usually plotted as log ν_{max} v. $1/T_{max}$ (frequency and temperature of the peak), thus obtaining a sort of activation energy for the relaxation process. Typical data of the frequency-temperature relationship for glass transition and for the secondary transition β of poly(methacrilate)s[35] are reported in Figure 11. We have deliberately exchanged the x-y axis of the plot with respect to those commonly used in literature in order to emphasize the variations expected for transition temperatures measured under different instrumental frequencies. It can be observed that it is impossible to unequivocally define a transition temperature without the associated frequency.

An apparent activation energy for the process can be obtained from the conventional diagram of log (ν_{max}) as a function of $1/T_{max}$ if it can be described by an Arrhenius-type behavior. In this case, the slope of the curves gives an apparent activation energy, $\Delta H^{\ddagger}$, that increases in passing from the low temperature to the high temperature transition. This reflects the increased size and molecular complexity of the domains involved in the different relaxations and therefore the increased value of the energies required for mobilization. We have used the adjective "apparent" to emphasize that the actual size of the molecular units involved in the different processes is unknown and therefore the quantities so far determined in these plots have only a comparative meaning.

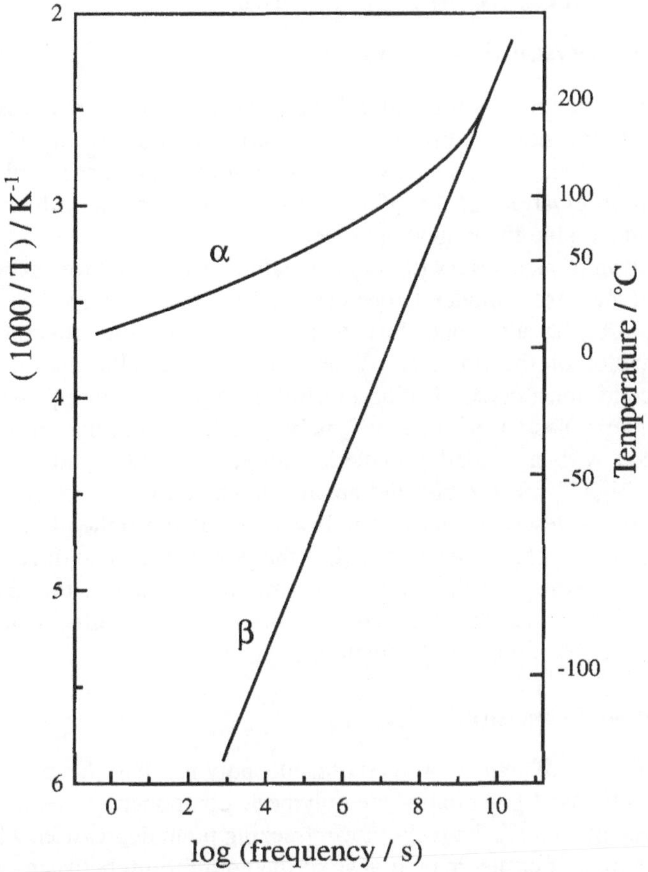

FIGURE 11 Dependence of transition temperatures for the relaxations α and β of poly(methacrylate) as a function of test frequency.[37]

Another observation concerns the convergence of transition temperatures of the two processes shown in Figure 11 at high frequency. At first sight, this trend suggests that low frequency measurements (thermodynamic limit) give a more resolved pattern of the transitions measured as a function of temperature, while it appears more difficult to attempt to measure the frequencies of the different transitions with a multifrequency apparatus by keeping the temperature constant. In some cases only one transition may occur. One may wonder whether, upon increasing measurement frequency, a secondary transition would move to a higher temperature across the glass transition. It has been claimed, on the basis of theoretical and experimental formulations, that in the limit of $1/T = 0$ (at infinite temperature) all the curves of $\log(\nu_{max})$ as a function of $1/T_{max}$ converge at the value of $\log(\nu_{max}) = 13$.[36] It seems reasonable to assume that a convergence of all the transition temperatures occurs at the value of the melting temperature (ca. 220°C).

EFFECT OF A SECOND MISCIBLE COMPONENT

DILUENT AND POLYMER PLASTICIZERS

In the previous section, a few rules have been provided for assessment of the dependence of transition temperatures, primarily T_m and T_g, upon structural or physical perturbations. We shall now apply these schematic rules in order to quantify the empirical observations of the effect of a second component (diluent, plasticizer) within certain suitable theoretical approaches.

The addition of plasticizers to a crystallizable polymer induces not only a strong T_g depression but also a moderate decrease in T_m. Consequently, the crystallization window ($T_m - T_g$) broadens and shifts to a lower temperature, affecting the crystallization behavior of the polymer. When a polymer is allowed to crystallize in isothermal conditions, crystallization kinetics change as a consequence of plasticization. Polymeric plasticizers are low T_g polymers that form a homogeneous mixture when blended with a second polymeric component with a higher T_g. Physical properties of high T_g macromolecules are affected by polymeric plasticizers in much the same way as described above for low molecular weight diluents. The main changes involve the glass transition and, if the polymer is crystallizable, also crystallization and melting. In the case of polymeric plasticizers, the glass transition broadens quite remarkably at intermediate compositions, the effect being commonly attributed to local composition fluctuations.

MELTING POINT DEPRESSION

The addition of a diluent to a crystallizable polymer has the primary effect of reducing the chemical potential of the polymeric component in the mixed solution. The most evident effect is the well-known freezing point depression, which is traced back to the simple thermodynamic laws of the equilibrium between the crystalline solid and the liquid solution to which a second component has been added. When dealing with cryoscopy, the melting point of water is lowered by the presence of a solute, the second component. Given the more complex structure of the polymeric phases, if the diluent is excluded from the crystalline lattice, the melting temperature of the crystalline fraction decreases with increasing diluent concentration according to the following equation, based on the Flory theory[37]:

$$\frac{1}{T_m} = \frac{1}{T_m^0} + \left(\frac{R}{\Delta H_u}\right)\left(\frac{V_u}{V_1}\right)(\phi_1 - \chi_1\phi_1^2) \quad (3)$$

where T_m and T_m^0 are the melting temperatures of plasticized and pure polymer, respectively, and ΔH_u, V_u, and V_1 are the molar melting enthalpy and the volumes of the polymer repeating unit and diluent, respectively. The concentration dependence is expressed by the volume fraction of the diluent, ϕ_1, and the polymer-diluent interaction parameter, χ_1.

The non-ideality is here expressed by the parameter χ_1, which is a function not only of the polymer/diluent species, but also of temperature and molecular weight.[40] The Flory theory assumes that the only change in the system morphology is dilution of the liquid polymer in the mixture. Therefore, all other conditions which affect the crystallization process, such as cooling rate or temperature of isothermal crystallization, will also influence the melting temperature of the polymer crystallized in the presence of a diluent. At constant temperature below T_m, the reduction of viscosity due to addition of a diluent may improve the diffusion-controlled growth of the crystals and partially counterbalance the decreased stability of the crystalline phase. In conclusion, we expect the addition of a diluent to modify the crystal melting point in a way which follows the Flory equation, but this effect is not cumulative.

Glass Transition Depression

The local friction coefficient in the glassy state is usually sharply reduced in a polymer system diluted with a low molecular weight solvent when a true solution is formed, in the sense that the solvent is molecularly dispersed. Each polymer chain segment has in its vicinity diluent molecules as well as other polymeric segments, and the former can be displaced in translatory motion much more easily, thus lowering the effective local viscosity. The resulting reduction in all relaxation times is the most striking effect on viscoelastic properties at constant temperature.[19] However, a transition temperature, such as the glass transition, will still be present as a discontinuity in the temperature domain. The total effects of a diluent or plasticizer on viscoelastic properties in the transition zone can, therefore, be visualized in terms of its influence on the temperature dependence function (Figure 9).

The addition of a low molecular weight diluent depresses T_g sharply and almost linearly at low concentration, as described in the following equation (note the resemblance to the melting point depression):

$$T_g = T_{g2} - k_{w_1} \tag{4}$$

where w_1 is the weight fraction of diluent and k is a constant which depends on diluent and polymer characteristics. The depression of T_g is attributed to the introduction of additional free volume with the diluent, as would be expected if the fractional free volume of the diluent (f1) exceeded that of the polymer (f2) and the free volumes were additive or approximately so. Since usually f1 > f2, this corresponds to $T_{g1} < T_{g2}$.

An extension of the linearity of the dependence of T_g over the whole composition can rarely be held as true, although this approximation is often assumed in order to predict the T_g of polymer blends.[22]

To give an overview of the various equations that explain the effect of a diluent in a mixture,[3,9] we should first distinguish between merely suitable and model-based equations. The first empirical description of changes in the T_g of a mixture was given by Gordon and Taylor[40]:

$$T_g = \frac{w_1 T_{g1} + k_{w_2} T_{g2}}{w_1 + k_{w_2}} + q_{w_1 w_2} \tag{5}$$

This equation (with addition of the empirical term $q_{w_1 w_2}$),[39] has been widely used in the context of polymers and other food-related systems (carbohydrates, proteins, etc.). The prediction, if any, requires determination by fitting of the empirical constant k. As we shall see below, more theoretical approaches provide a molecular description of the value of k, although this does not imply that the theory is correct.

A more complete formulation of changes in the glass transition of a polymer-diluent mixture can be derived from the thermodynamic theory formulated by Couchman and Karasz[41,42] for binary blends of miscible polymers. This theory is based on the continuity conditions of entropy and volume functions at the glass transition temperature for a miscible two- (or more) component system. The theory is not predictive of whether a single T_g is observed, since compatibility of the two components is assumed. The equation formulated by these authors would imply knowledge of either the heat capacity of the two components or (in the case of the thermodynamic equilibrium, as it is here) the isobaric volume expansivity. Unfortunately, the exact logarithmic equation has not been used with great enthusiasm by practitioners, and the simpler, more approximate equation (also provided by the authors in the original paper) is commonly quoted:

$$\ln \frac{T_g}{T_{g1}} = \frac{x_2 \Delta C_{p2} \ln(T_{g2}/T_{g1})}{x_1 \Delta C_{p1} + x_2 \Delta C_{p2}} \tag{6}$$

This equation allows us to identify the constant k of the empirical Gordon and Taylor equation as equal to the ratio of heat capacity changes of the two components; that is, $k = \Delta C_{p2}/\Delta C_{p1}$. This fact lends importance to the calorimetric studies of the glass transition, not only for determination of transition temperature, but also because knowledge of the heat capacity steps provides a quantitative description of the composition dependence of the transition temperature. The linear approximation of the Couchman-Karasz equation as reported above is identical to that derived by Gordon et al.[40] from the Gibbs-Di Marzio theory of glass transition.

Reliability of the T_G-Composition Curves

The results of all these equations are clearly different, and the deviations in temperature from that obtained from the simplest linear combination can be as large as 50°C. Figure 12 is a reconstruction of the curves predicted by Equations 2 and 6, the simple Fox equation and the theoretically founded Couchman-Karasz equation. The simulated system could be, for example, water-sucrose, for which experimental data exist in literature.[43,44] In the sugar context, however, it is important to mention that even for determination of the T_g values of simple sugars, a number of spread data exist in literature.[39,45] These uncertainties need accurate screening of existing experimental data and critical evaluation of experimental conditions before their values can be accepted as reference data.

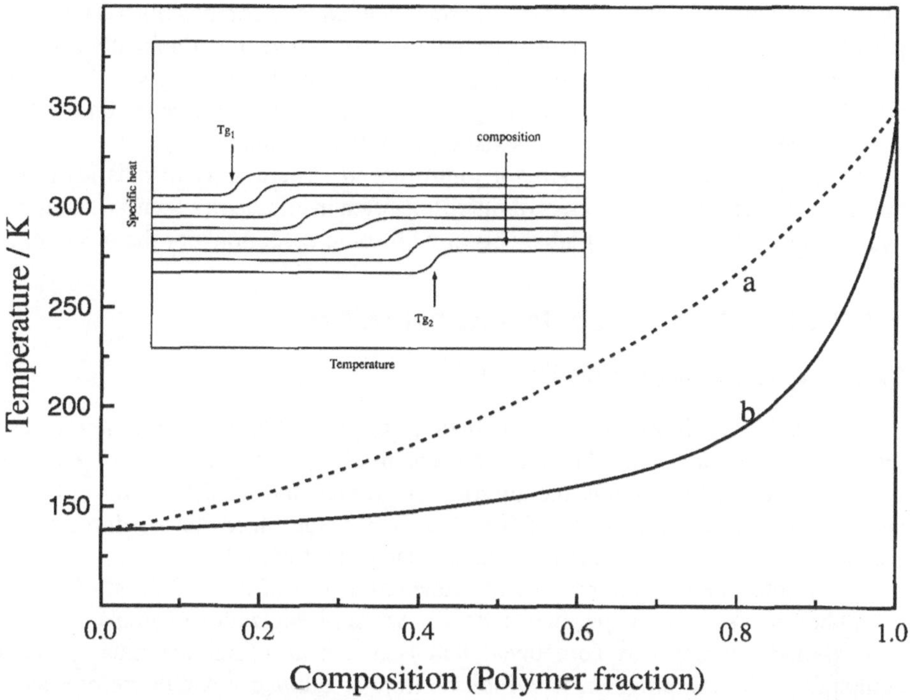

FIGURE 12 Effect of composition on glass transition temperature. Line a is calculated with Equation 2 and line b with Equation 6. The insert shows the broadening of the heat capacity step which arises from incomplete miscibility, producing a cusp in the plot (not shown).

The evidence and prediction of the effect of addition of small amounts of plasticizer in cross-linked systems is of great interest. ten Brinke et al.[46] have shown an enhanced sensitivity of depression of the T_g of polystyrene, when cross-links are introduced as a result of adding divinylbenzene as a co-monomer. The initial slope of the T_g curve (dT_g/dx_2) changes from 650K to about 1000K for a co-monomer composition of 35.7%. The authors also offered a classical thermodynamic theory based on the Couchman-Karasz approach to obtain expressions which satisfactorily predict the experimental values.

A final warning must be added about any attempt to draw a curve based on a few scattered data of T_g over the whole composition, even in the absence of knowledge of whether the system is completely miscible. It is possible that the occurrence of a cusp[47] or a singularity associated with two different composition dependencies of T_g at low and high plasticizer content may greatly affect the reliability of the interpolated values (Figure 12). A singularity in the T_g-composition dependence of plasticized polymers has been predicted in literature, and two theoretical treatments are available. The earlier one is based on free volume considerations, and the later one on purely thermodynamic grounds. According to these treatments, in polymer-plasticizer mixtures below a certain critical temperature, the polymer contribution to T_g either vanishes or remains constant, and a cusp is generated in the T_g-composition dependence. In all cases where such anomalous behaviors were reported, broad glass transitions were observed in the intermediate range of compositions. Therefore, plasticized polymers should not be expected to show a single monotonically decreasing glass transition over the entire compositional range. Rather, they should exhibit two distinct or partially overlapping glass transitions with different composition dependence associated with two coexisting mobilization phenomena.

Many other questions are left for future debate. The quality of data about the bread system and the biopolymers of interest is often insufficiently accurate to enable us to evaluate the analysis reported above, with some exceptions for the recent data.

WATER PLASTICIZATION IN BIOPOLYMERS

RATIONALE FOR COMPOSITION DEPENDENCE

To look at the plasticization of biopolymers, examples should discuss the effect of the presence of water and illustrate the problem of the presence of two T_gs due to phase separation. In addition, the question of average and local behavior in binary and ternary systems (with and without phase separation) needs to be addressed for the case of addition of a second low-molecular weight consolute.

We would like to reinforce one fundamental aspect of the definition of a plasticizer before we discuss experimental data on water plasticization of simple biopolymers (binary systems). A component that is added as plasticizer must be fully compatible with the polymer, therefore, no immiscibility gap should be present in the phase diagram of a plasticizer/polymer system, and no other phase-separation phenomena should occur.[48] These phenomena do not always need to be absent from experimental cases, but where such a discontinuity occurs, the effect of the plasticizer is interrupted. In the presence of two phases (either mascroscopically or microscopically separated), each phase will exhibit its own properties. Lack of this fundamental key to interpretation has, in many instances, given rise to misconceptions and misleading descriptions. Once this test for the correct process is used, a simpler generalization is possible. Similarly, acknowledging that ionic dissociation of electrolytes takes place enables the thermodynamic formalism to describe the physiochemical properties of an otherwise bizarre class of solutes.

The fundamental question is whether it is possible to fully explore the composition domain of a system made by solvent and polymer. It probably is possible for only a limited number of cases, none of which includes the bread polymers/water systems. An example of a realistic phase diagram and the consequences of phase separation, the schematic behavior of a mixture of solvent/polymer under different phase separating conditions, is shown in Figure 13.

MISCIBLE SYSTEM: LOW MOISTURE REGIME, LOW POLYMER CONCENTRATION REGIME

In Figure 13 a critical concentration can be identified which separates two different temperature-dependent phenomena. Starting with a pure polymer, to which increasing amounts of diluent water (low moisture content) are added, the T_g of the polymer-water system decreases with increasing water content, according to the relationship

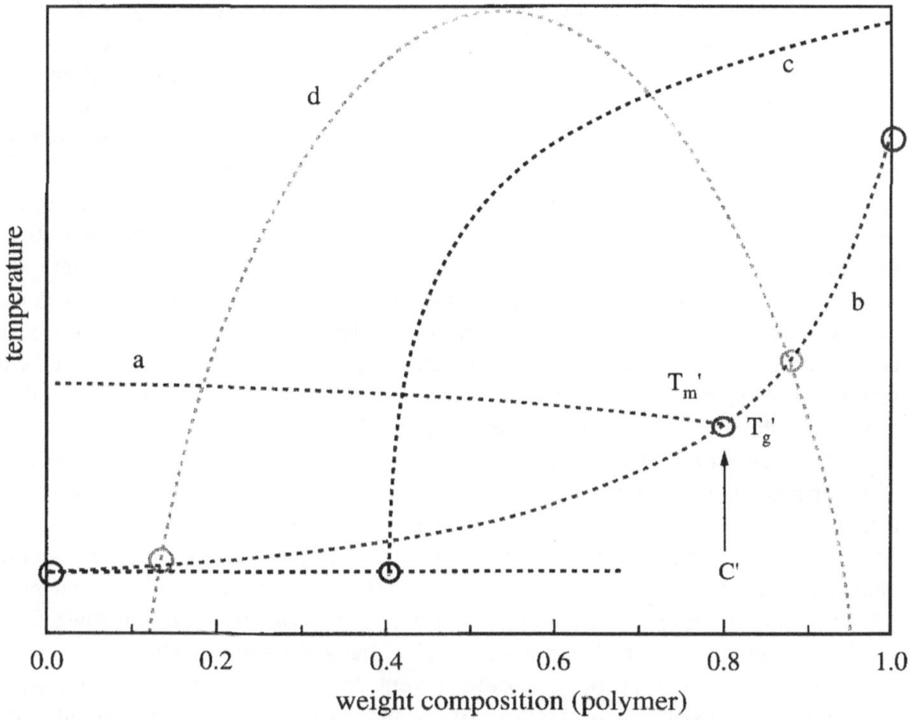

FIGURE 13 The interplay of phase demixing on the apparent melting and glass transition temperatures. The lines are: a) solid-liquid equilibrium line for the solvent; b) glass transition line for the plasticized polymer; c) solid-liquid (equilibrium) line for the crystalline polymer, d) hypothetical line of an immiscibility gap. Note that line c) has been drawn by taking into account the changes predicted by both Flory-Manderkern and Hoffmann-Weeks equations.

illustrated previously (the Couchman-Karasz equation). In principle, provided that the procedures for obtaining the glassy state are followed (i.e., fast quenching), then the value of the T_g measured would decrease steadily, reaching the final value of the T_g of the pure solvent water at about –135°C. No other state would be obtained on cooling apart from the glassy one. The existence of the glassy state is usually revealed on heating the system. The typical DSC thermogram shows a glass transition produced by the amorphous phase. If the amount of water is higher than C′, this transition can be followed by a cold crystallization of the water at temperatures above T_g but below T_m. The solid ice formed will then melt at a melting point T_m in equilibrium with the liquid solution (water-solute).

The increasing water content usually introduces enhancement of molecular mobility which can still be sufficiently high when crystallization of the solvent is started at low temperature, thus preventing the formation of a glassy state which would occur at lower temperature. With formation of ice crystals, the mixed water-polymer phase becomes progressively richer in polymer, and the $T_c - T_m$ of the water/ice equilibrium always shifts to a lower temperature, eventually reaching the value of T_m'. The uniqueness of this point T_m' is illustrated below.[13,39] Assuming that no undercooling effects are present for the solvent ($T_m = T_c$), the value of T_m varies with concentration according to the thermodynamics of real systems, i.e., changes in the activity of the liquid solvent water where temperature and composition must be taken into account. The full line has been drawn to schematically follow this thermodynamic behavior of water mixed with a real hydrophilic solute (e.g., sucrose, for which abundant, though scattered, data are available in literature).

On slowly cooling a water-rich solution, the concentration of the solute increases due to the crystallization of ice (freeze-concentration). Both the water content and the temperature decrease, reaching the intercept T_g'/C', where the equilibrium melting/crystallization curve crosses the glass transition curve. At this point, two phases coexist, one pure crystallized water, the other the glassy state of the mixture at composition C′. On heating, at the temperature T_g' the phase of composition C′ will result in transition to the amorphous state, which will be progressively diluted by melting of the ice as the temperature increases above T_g'.

Reheating an effective glassy state rich in water appears complicated because of the irreversible process of cold-crystallization of the solvent, which is typical of a metastable state. However, as in a pure quenched polymer, crystallization may occur to a limited extent due to limitations imposed by thermodynamic and kinetic considerations. Crystallization follows the rules given previously and is strongly influenced by the rate of diffusion processes and the potential gradient towards an equilibrium condition which, upon heating, is changed continuously.

At low water content, ($C < C'$) the solvent does not crystallize because when the temperature is decreased, the viscosity of the system increases dramatically and the glass transition is passed through before any crystallization can be achieved.

Glass Transition of Starch and Protein Systems

Determination of the glass transition temperature of starch polysaccharides has not been an easy task. The T_g of the whole starch has been the subject of many controversial reports. This is not surprising in view of the fact that even for a simple and well characterized sugar such as sucrose the literature values range between 50 and 70°C.[45] Table 3 contains some values for starch polymers as reported in literature. It is not the scope of this chapter to review all the literature data; however, it is difficult to extract what should be the most reliable value of T_g for a defined

TABLE 3
Glass Transition Temperature (°C) of Starch Components

Nakamura and Tobolsky, 1967[49]	330	Anhydrous amylose, extrapolated
van den Berg, 1981[50]	151	Anhydrous starch, extrapolated
Zeleznak and Hoseney, 1987	—	Water content 13–22%
Orford et al., 1989[29]	227 ± 10	Linear polymer, extrapolated *vs.* M
Whittam et al., 1990[52]	197	Anhydrous starch, extrapolated
Roos and Karel, 1991[53]	243	Data on maltodextrins[29]
Bizot et al., 1997[54]	300–330	Linear polymer, cast

component from the above data because the molecular composition of the system is not always given. A higher value is expected for the linear component and would, therefore, suggest that the best current values for the anhydrous linear polysaccharide, amylose, is probably above 300°C. Bizot et al.[54] reported a circumstantial analysis of the factors which influence the T_g of starch polysaccharides (i.e., molecular weight, chain branching, chain flexibility, crystallinity, and blending), and they conclude that "all the trends seem orthodox from a polymer science point of view," giving strength to otherwise pedantic concepts.

The T_g of starch can be as high as 300°C, while other transitions that are commonly attributed to melting occur at lower temperatures. However, melting endotherms[55-58] are known to be strongly affected by the presence of plasticizers. Glass transition temperature in a mixed polymer system is a function of the composition according to the Gordon-Taylor equation, while melting temperature depends to a limited extent on the composition, the morphology, and the size of the crystals. Therefore, the temperature of melting at molecular size can decrease much more than the glassy transition temperature. The rule T_m/T_g = constant applies only to macroscopic amorphous and crystalline phases.

At molecular level, the melting transition between ordered species such as globular proteins, which can be assimilated into very small crystals, and the disordered chain is a pseudo-first order transition.[59] The T_m of this species eventually becomes constant and independent of the concentration. The proteins present in foods are not always globular from the conformational point of view, and structural organization ranges from the triple-helical domains of gelatin to the essentially disordered conformation of some dough proteins. Still, some regularities have been found which show the T_m of the molecular domains (in dilute solution) related to T_g of the proteins.[60]

Effect of a Second Soluble Low-Molecular Weight Component

Many results indicate that the plasticizing effect of water, the factor considered most important in literature, is often decreased by the presence of small amounts of a second plasticizer. This effect is ascribed to reduction of the quantity of water available for efficient plasticization. However, careful scrutiny of a tridimensional ternary diagram of the T_gs surface (Figure 14) clearly illustrates a simpler explanation for the effect on the T_g of the mixture upon the addition of a low molecular weight solute. Different behaviors can be obtained depending on the composition of the polymer/water mixture. Decrease, constancy, or increase of T_g value is obtained depending on height of the water content. The partial Gordon-Taylor curve of the ternary system along the coordinates with constant polymer-water composition illustrates this. Complicated speculation has been offered in the literature, claiming hydrogen bonds, molecular compactness, etc. Unless clear evidence is given by spectroscopic methods these speculations must be discarded. The occurrence of a immiscibility gap, although it does not invalidate the above, must be taken into account in terms of a failure in the continuity of the Gordon-Taylor behavior.

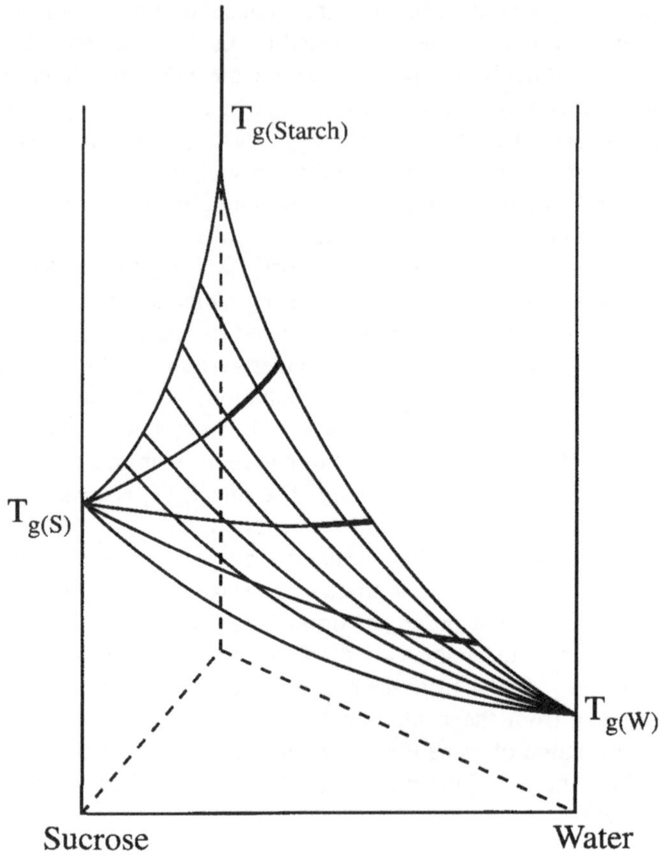

FIGURE 14 Ternary diagram of the glass transition temperature, T_g, for a mixture of three miscible components: polymer (starch), consolute (sucrose, S), and water (W). Three bolded segments show the different changes (negative, almost zero, and positive) of T_g of the polymer-water system at different composition by the addition of the third component (S).

The effect of the solute addition to a ternary water/polymer1/polymer2 system may be more complicated if the solute is compatible with only one polymer. This may be the case of massive additions of sucrose to a water/protein/polysaccharide mixture. For our purpose, although gluten is not water soluble, it is clearly water plasticizable, and both amylopectin and gluten are substantially plasticized by water but sugar behaves differently in these two systems. Various factors may cause the reduced plasticizing (T_g reducing) effect of sugars on gluten when compared with amylopectin. In the case of amylopectin, sugars are very similar to or identical to starch monomers and as such are readily incorporated into the system at low sugar contents, with a resulting reduction in T_g due to the relatively high mobility of the sugar molecule. In the case of gluten it may be more difficult to attain uniform mixing due to incompatibility of the sugar, resulting in a reduced effect on T_g even when the method of sample preparation is changed. The presence of incomplete compatibility is shown by broadening transition region as observed in the presence of sugar and the appearance of a transition in the region of the sugar T_g.[61-63] Glycerol appears to be more miscible with gluten due to its lower molecular weight. Another possible reason for the reduced plasticizing effect, or even T_g increasing effect of sugars, is that sugar molecules may be preferentially hydrated, decreasing water content and therefore increasing T_g of the gluten matrix. Sugars have also been observed to have a stabilizing effect on native conformations of proteins, because of their thermodynamic incompatibility with the unfolded chains, which would also tend to increase the T_g of the system. Thus, several effects make it difficult to predict the T_g of these three component systems.

Effect on the T_M

The melting process is associated with structural changes at a large scale. The effect of the plasticizer can be exerted on the ordered or melted phases. The presence of water or another component is believed to affect mainly the disordered phase of a biopolymer, unless the small molecules are cocrystallized or they intercalate and take part of the crystalline architecture. It is often possible to distinguish from the shape of the DSC curve of crystalline phases whether the transition is a cooperative process between well-defined states or if several ordered states with different stability are present. The sharpness and symmetry of the bell-like DSC curves is indicative of a cooperative process between well-defined states. Consider a microdomain of a structured biopolymer, such as a globular protein. The phase transition (denaturation) of a native protein in aqueous solution is subjected to several perturbations which may arise from interactions between protein and other molecules in solution. However, the size of the thermodynamic domain remains unaltered and a constant melting temperature is obtained if the stability of the unfolded species is not changed due to the presence of consolutes.

Whenever large changes in the temperature and in the enthalpy of transition are observed, this is associated with large structural changes. Donovan[55] reported that at a high water level only one endotherm at 66°C is observed. As the amount of water is decreased, a trailing shoulder develops which, at a sufficiently lower water level, becomes the only endotherm present. He concluded that the high-temperature phase transition is a melting of crystallites and the dependence of the temperature of this endotherm on water content can be treated thermodynamically. The conclusion has been reached over the years that the endotherm around 60°C is ascribed to the melting of the amylopectin double helices or similar structures, while the higher temperature peak behaves like that of crystallites of variable size.

BINARY AND TERNARY SYSTEMS WITH PHASE SEPARATION

Phase Separation: One or Two T_gs?

In 1896, Beeijerinck[5] reported that two aqueous solutions of biopolymers cannot mix together (for example, gelatin and starch). Similarly, Sperling states that two polymeric solutions are generally incompatible, but he uses a distinct nomenclature for the different thermodynamic aspects of phase separation.

The phase separation of gliadin and glutenin fractions of wheat flour from water-soluble proteins, starch, and pentosans seems to contribute to some reproducibility of the physiochemical behavior of wheat flour components. As a buffering action, this behavior minimizes the changes resulting from a modification of the composition of the dough proteins and the aqueous medium. At least two aqueous phases containing proteins are formed in the dough. The first is the concentrated protein viscoelastic phase containing gliadins and glutenins, known as gluten. The second coexisting phase is a viscous mixed solution of albumins, globulins, neutral and charged polysaccharides. The existence of these two phases has partially modified the previous description of gluten as a protein complex. The term gluten refers to a highly concentrated protein phase with the characteristics of a thixotropic gel phase. The other protein-polysaccharide viscous phase may be treated as a liquid (concentrated) phase. There are at least four different dough phases (liquid, pentosan gel, gluten, and starch) which cannot be separated by ultra-centrifugation[64] at water content below *ca.* 35%. The ease of washing starch out of wheat flour dough reflects a low affinity between starch granules and the gluten phase. However, this process is a separation of phases but not a phase separation.

Evidence has been provided for an unusual T_g-concentration dependence for the plasticization of synthetic polymers.[65] The conclusion was that this effect results from a phase separation, and one of the phases has to be either pure polymer or a very concentrated solution, whereas in the second phase vitrification causes the anomalous T_g. It is now widely accepted that miscible polymers have a single T_g between the T_gs of the two components and a broadening of the transition as the components become more immiscible, but an incompatible system will have two T_gs corresponding to the two components. However, if the component T_gs are less than 20°C apart, this method does not enable us to distinguish a phase-separated system from a miscible one. This situation has always been found in biopolymer systems.

Another example can be seen in the effect of water on the glass transition of 1:1 mixtures of amylopectin, casein, and gluten, as studied by Kalichevsky et al.[66] The problem is also discussed below in terms of the effect of adding another component to structural reorganization as, for example, during crystallization.

Plasticization Effects on Eterophasic Structural Reorganization

In either miscible or non-miscible systems, polymer components undergo reorganization processes which depend on the mobility of the segmental elements and on the energy barrier for formation of critical size nuclei. Even in the presence of a demixing process, the growing rate of the structural organization is believed to be the same as that in the pure component, provided phase separation is complete. The simplest hypothesis that can be made is that the two polymers are immiscible over a suitable temperature range. This condition corresponds almost to the case of the polysaccharide-protein aqueous system. However, the morphological aspects of crystallizable components in the phase-separated systems is of great importance for the final texture of the product.

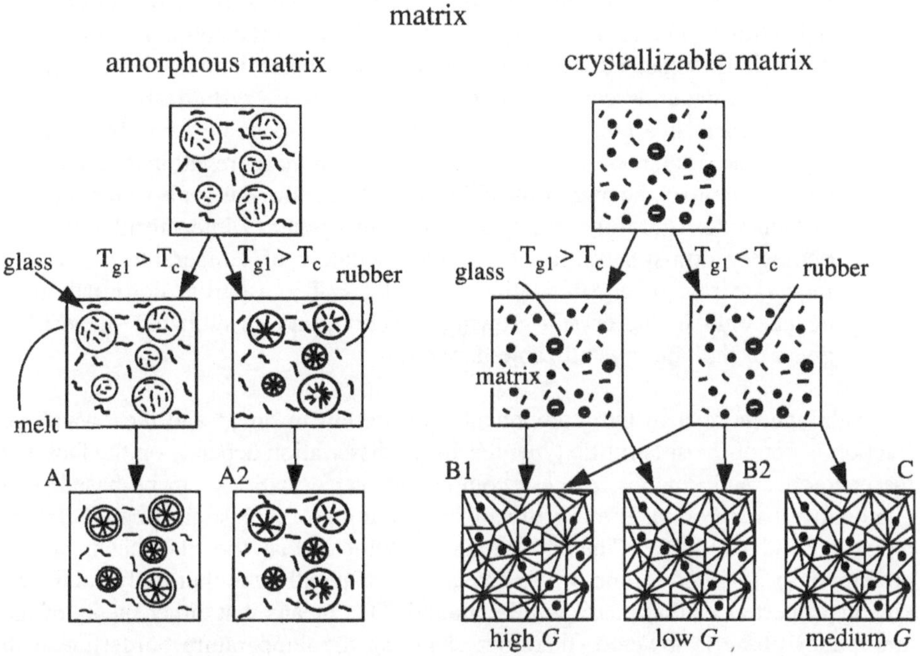

FIGURE 15 Structural reorganization in incompatible polymeric mixtures. Case A: the crystallizable polymer is surrounded by a non-crystallizable amorphous matrix (below T_{g1} case A^1 or above T_{g1} case A^2). Cases B and C: the polymeric matrix is crystallizable. The texture depends on the temperature of crystallization and growing rate (G) of the crystalline polymer.

If partial miscibility has to be taken into account, the two phases in equilibrium with different compositions can be treated in an analogous way. The phase rich in the crystallizable component gives faster and more extensive crystallization, while the conjugated phase remains similar to an amorphous immiscible phase. The morphology of the final structure depends on whether the phase-separation process is faster than the crystallization process.

Two distinct cases (Figure 15) are identified on the basis of whether the continuous phase is made up of the non-crystallizable component (polymer 1) or the crystallizable component (polymer 2). In addition, each case presents two possibilities, since the crystallization temperature T_c can be lower ($T_{g1} > T_c$) or higher ($T_{g1} < T_c$) than the glass transition temperature of the amorphous component.

1. When the continuous matrix is amorphous and $T_{g1} > T_c$, crystallization of the molten domains occurs within a non-deformable glassy matrix, and the volume contraction on cooling generates holes in the inter-phase. The mechanical properties will, therefore, be poor. If $T_{g1} < T_c$, the rubbery matrix is adaptable to the volume changes which occur during crystallization, and the final structure is more compact.
2. When the continuous phase is made up of the crystallizable component, the dominant factor is the ratio between rate of spherulitic growth and

> rate of deformation of the non-crystallizable domains. If the rate of crystallization is high (either for $T_{gl} > T_c$ or for $T_{gl} < T_c$), the non-crystallizable phase is interspersed within the crystalline superstructures in the same way as in the undercooled liquid (B1). However, if crystallization occurs slowly, there is a tendency to exclude the amorphous domains from the crystalline front, and eventually an inter-spherulitic segregation is reached (B2). Of course, the migration of the amorphous domains is also a function of their size. Therefore, segregation may also generate inter-fibrillar morphology (intra-spherulitic). In a small window of temperature with a moderate rate of crystallization (with $T_{gl} < T_c$), a partial deformation occurs only in the crystal-growing direction, and an ellipsoid shape is given to the non-crystallizable domains.

Although more than two components are present in bread and the crystalline fraction is not of the spherulitic type, the above dissertation depends on the fact that the energetics and kinetics of these complex phase formations can be based on a quantitative background, and the effect of the addition of a plasticizer can therefore be better understood.[67,68] The most relevant point is that the plasticizer can be preferentially solvated in one of the phases. Consequently, the thermodynamic and kinetic properties of that phase will be altered. This means that either the T_g of the amorphous phase is lowered (therefore changing the temperature borderline with the T_c) or presence of the solvent in the crystallizable phase affects melting temperature (i.e., the undercooling necessary for crystallization) and the growth rate of crystals. In all cases, segregation of the plasticizer occurs if the crystalline phase excludes the small molecules from the crystalline lattice. The diluent, therefore, becomes an effective third interstitial phase. A partial summary of the plasticization effect on bread components is shown in Figure 16.

PLASTICIZATION AND PHASE MOBILITY IN BREAD

THE POLYMER CONCEPT OF SOFTENING AND TOUGHENING

Let us summarize some concepts about the mechanisms of softening and toughening polymers at the micro-morphological and mechanical level. Assuring good property response to the engineering plastics is an important task. Major properties of softened-toughened materials, as outlined below, apply to synthetic polymers as well as bread polymers.

1. The glass transition temperature should be at least 60°C below ambient. For most practical aspects, this means that the T_g of a food material should be below −40°C.
2. The domain size of the micro-structures must be of the order of a fraction of a micron. For many materials, the optimum equivalent diameter seems to be about 0.3 μm.
3. The spacing between structured particles should be of the order of some μm. This distance is considered sufficient for reacceleration of the crazing

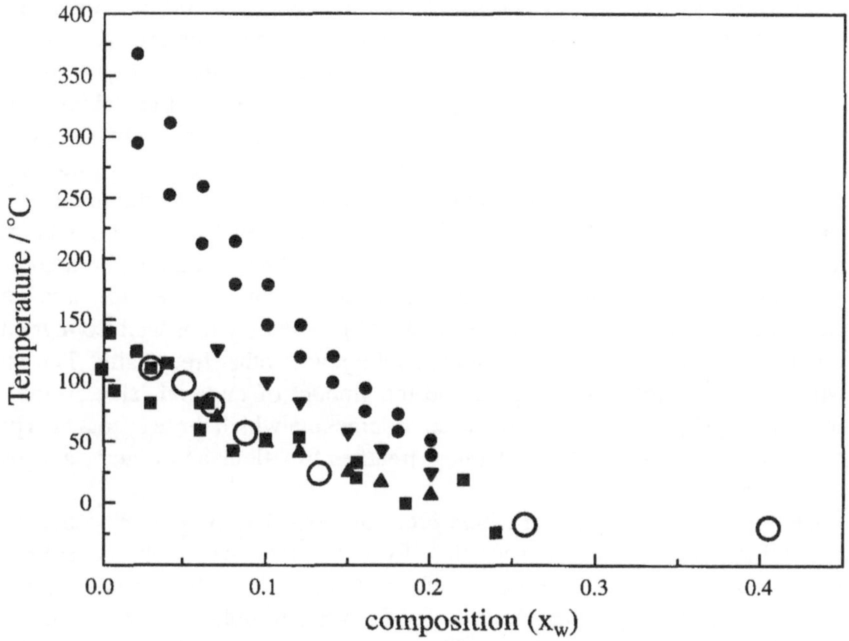

FIGURE 16 Effect of water content X_W (w/w) on the T_g of bread and its components. Data have been redrawn.[54,61,66,69-71] Starch amyloses (full circles) have higher T_g than gluten polymers (full squares). A mixture of gluten-amylopectin gives two T_gs (upper and lower triangles). White bread T_g values are open circles.

or shearing mechanism before the next particle is encountered. In polymer composite materials, the micro-failure mechanism may change if the distance between particles is too large, while proper particle spacing enhances cooperative cavitation (i.e., the formation of vapor-filled cavities between solid boundaries).

4. Since most types of elastomeric components (as gluten) are cross-linked, care should be taken to achieve high enough cross-linking so that the particles do not deform unduly during processing, but are able to cavitate under imposed stresses at a later stage. In addition, there must be good adhesion between elastomeric particles and plastic components.

Once again, the relationship between end-use properties, as well as the modifications that other components might introduce, and the characteristic transition temperatures of the materials is evident. Some of these concepts are related to structural re-organization in a multicomponent system and find a direct application in the following description of structural changes in the presence of the plasticizing water.

The Process of Gelatinization

As starch is the crystallizable component of bread, it is reasonable on the basis of weight to say that starch plays a significant role in toughening. However, the roles of all polymers within the framework of their ability and availability to interact with water and manifest relevant physical properties should be taken into account.[72]

Native starch granules in flour are water-insoluble, partially crystalline materials that reversibly swell in the dough stage, albeit only slightly. After baking, microscopic observation of the crumb shows that most granules are partially swollen and separated by a thin layer of gluten protein. These changes occur during baking, whereby the combination of heat, moisture, and time transforms the granules into an amorphous (non-crystalline) state. On a molecular level, associations between starch molecules are relaxed, granules gelatinize and swell, and a small amount of starch is leached into the inter-granular phase. At granular surfaces, the outer chain branches of the large amylopectin molecules are presumably loosened from granular restraints sufficiently to reassociate, upon cooling, with other free chains. The extent to which granular order is disrupted and the amount of material released into the inter-granular gel phase depends on the environment in which these events take place (temperature, starch content, mechanical treatment of flour and dough, and water level and availability).

Gelation is the phenomenon which produces solid-like properties (toughening) in a liquid system or, more appropriately for our cases, gives mobility (softening) to a solid-like material. The synergistic effect of gelling biopolymers is a significant manifestation of excluded volume effect of food macromolecules, which are often characterized by semi-rigid chains.[73] The most simple description of the polymer gel system requires that "at least three different chains converge in the junctions."[77] No limitations have to be assumed a priori on the length of the flexible spacers between junctions or the size or shape of segments entering the junctions.

The common definition of polymeric gels is given for cross-linked chains and excludes the reversible gels which are important in food systems.[8] Most of these gels are thermo-reversible, while some of them (e.g., alginates and carrageenans) are controlled by chemically reversible ionic interactions. In all cases, gelation is thought to take place in a fluid system upon the generation of either chemical or physical cross-links which give rise to solid-like properties. The gel-point (temperature and concentration) is the point where, in viscoelastic experiments, the elastic component G' becomes larger than the viscous component G''. However, the term gelatinization has been used in the context of bread to mean the process during which crystallites partially melt and dissolve in the starch system, giving the appearance of a gel. It is precisely during the gelatinization of starch that conversion of the starch granules takes place from solid to liquid droplets, and then may produce a solid filler of the gluten matrix. The rate of starch recrystallization and the glass transition temperature of starch strongly depend on the moisture level. Relaxations below the glass transition temperature have also been found in bread. The intensity of the relaxations is a function of added plasticizers, water, and consolutes.[75,76] Remarkable differences in texture and staling rates between different breads can easily be ascribed to different water content (e.g., pan breads and Chinese steamed breads). The liquid-protein phases of dough could also act as a filler of the gluten phase. Extraction of the water-soluble components of the dough, therefore, results in a considerable increase in the elasticity modulus of the gluten phase. Baking is accompanied by water evaporation, denaturation of proteins, and starch gelatinization. Because starch granules are a filler of both the liquid and gel-like protein phases of dough, starch gelatinization may provoke de-watering of both dough phases. Glass transition of the gluten phase within the exterior layer of the loaf fixes its structure, shape and volume, and retards staling.

A recent study of the organization of starch granules may help us to better understand these complex processes. Although the study was obtained for potato starches, it nevertheless outlines the astonishing molecular organization of several distinct phases.[77,79]

Intra-phase Mobility of Water Plasticization

Volume is an important property for characterization of the dough phase. However, volume changes and enthalpy changes occurring during phase transitions are much smaller than changes in the mechanical properties. Expansion and softening are not always related. Onset of expansion is much less affected by changes in hydration than is softening. Therefore, softening is considered a more satisfactory indicator of the glass transition of the cell walls than are volume changes.[70] It has also been established that the softness of breads with pentosans can be attributed to a higher moisture content rather than to a mere effect of these gums, since it is well known that higher bread moistures promote crumb softness.

Moisture affects the development of the crystal structure of starch components. Experiments have shown that breads held in a 98% RH to ensure a constant moisture level throughout the storage period are soft and acceptable. Overall crumb moisture is maintained, demonstrating the efficacy of moisture for softening the crumb by acting as a plasticizer. Conversely, bread sealed in a plastic pouch and aged in its own atmosphere leads to a harsh and unacceptable crumb. X-ray diffraction shows a medium diffraction intensity, indicating a degree of hydration resulting only from moisture migration into the crystalline regions formed during storage. The process of moisture transfer substantially affects all the physical and perceptive properties of bread. The most feasible source of such moisture would be the amorphous fraction in the crumb, resulting in a dried crumb and a bread harsh in texture and unacceptable.[80]

Water is recognized as an efficient plasticizer in the amorphous regions of starch but seems less active in the unacceptable bread. Loss of moisture from gluten to starch and the water-soluble components has been suggested. However, the amount of water that gelatinized starch equilibrated at an atmosphere of *ca.* 98% RH can hold depends on the age of starch gel. Instead, the water-sorption capacity of the baked gluten changes very slowly, if at all, over seven days. Any moisture transfer, therefore, would take place from starch to gluten. The progressive drop in starch moisture-sorption capacity and the absence of change for gluten is an important point to observe. At any particular time an equilibrium will exist for the distribution of moisture among the constituents of the bread crumb. The slow release of crystalline-bound hydrate water, which takes place during the local process of syneresis, affects crumb plasticity and influences staling perception, and is therefore effective in extending bread shelf life.

FINAL REMARKS

The framework of the arguments which have been used for this chapter have been developed on the basis of original articles in the field of polymers and polymer textbooks, with minimum reference to published reviews on plasticization in foods. Reference to the many reviews in specialist journals have been avoided. In particular, the authors are grateful for the generous list of references which can be found in the various articles by H. Levine and L. Slade.[81] It would be a Herculean task to condense many thousand pages into a few concepts and basic descriptions.

Major factors in understanding plasticization are awareness of the "common behavior" as well as situations which may appear to contradict the rules. Only in this way can the "know-how" be correctly replaced by the "know-why". The goal is the most effective transfer of knowledge from "polymer science and technology" to "food polymer science". Natural macromolecules were studied well before synthetic polymers, and the polymer industry used natural sources before the synthetic rubbers.[82] Staudinger's struggle for macromolecules is an example of the difficulty of acceptance of an unusual class of compounds and their peculiar properties. While progress in the field of synthetic polymers has been supported by industrial interests, the field of food polymers has received much less support.

ACKNOWLEDGMENT

The paper has been prepared with financial support of M.U.R.S.T. and University of Trieste. The helpful interactions with partners of the projects of European Commission FAIR CT 96-1015 and FAIR CT 97-3609 are also acknowledged.

REFERENCES

1. Haken, H., *Synergetics*, Springer Verlag, Berlin, 1978.
2. Sperling, L. H., *Polymeric Multicomponent Materials*, John Wiley & Sons, New York, 1997.
3. Davidou, S., LeMeste, M., Debever, E., and Bekaert, D., Contribution to the study of staling of white bread: effect of water hydrocolloid, *Food Hydrocolloids,* 10, 375, 1996.
4. Piazza, L. and Masi, P., Moisture redistribution throughout the bread loaf during staling and its effect on mechanical properties, *Cereal Chem.,* 72, 320, 1995.
5. Tolstoguzov, V., Thermodynamic aspects of dough formation and functionality, *Food Hydrocolloids*, 11, 181, 1997.
6. Marangoni, M., *Saper vedere,* Garzanti, Milano, 1953.
7. Caglioti, G., *The Dynamics of Ambiguity,* Springer Verlag, Berlin, 1992.
8. Alger, M. S. M., *Polymer Science Dictionary,* Elsevier Applied Science, Barking, 1989.
9. Johari, G., Pascheto, W., and Jones, S. J., Intergranular liquid in solids and premelting of ice, *J. Chem. Phys.,* 100, 4548, 1994.
10. Hoffmann, J. D., Davis, G. T., and Lauritzen Jr., J. I., Crystallline and noncrystalline solids, in *Treatise on Solid State Chemistry,* Vol. 3, Hannay, N. B., Ed., Plenum Press, New York, 1976, chap. 7.
11. Hiemenz, P. C., *Polymer Chemistry,* Marcel Dekker, New York, 1984.
12. Strobl, G., *The Physics of Polymers,* Springer-Verlag, Berlin, 1997.

13. Franks, F., Water and aqueous solutions at subzero temperatures, in *Water, a comprehensive treatise*, Vol. 7, Franks, F., Ed., Plenum Press, New York, 1982, chap. 3.
14. Amor, S. R., Rayment, T., and Sanders, J. K. M., Polydroxybutyrate *in vivo*: NMR and x-ray diffraction characterisation of the elastomeric state, *Macromolecules*, 24, 4583, 1991.
15. Keller, A., Goldbeck-Wood, G., and Hikosaka, M., Polymer crystallization: survey and new trends with wider implications for phase transformations, *Faraday Discuss.*, 95, 109, 1993.
16. Hutchinson, J. M., Physical aging of polymers, *Prog. Polym. Sci.*, 20, 703, 1995.
17. Eyring, H. and John, M. S., *Significant Liquid Structure*, John Wiley & Sons, New York, 1969.
18. Doolittle, A. K., Studies in Newtonian flow. II. The dependence of the viscosity of liquids on free space, *J. Appl. Phys.*, 1471, 1951.
19. Ferry, J. D., *Viscoelastic Properties of Polymers*, John Wiley & Sons, New York, 1980.
20. Gibbs, J. H. and DiMarzio, E. A., Nature of the glass transition and the glassy state, *J. Chem. Phys.*, 28, 373, 1958.
21. Bondi, A. and Tobolsky, A. V., *Polymeric Science and Materials*, Tobolsky A. V. and Mark, H. F., Eds., John Wiley & Sons, New York, 1971, chap. 6.
22. Sperling, L. H., *Introduction to Physical Polymer Science*, 2nd ed., John Wiley & Sons, New York, 1992.
23. Cowie, J. M. G. and Henshall, S. A. E., The influence of chain length and branching on the transition temperature of some polyglucosans, *Eur. Polym. J.*, 12, 215, 1976.
24. Chu, B., McWherter, Brant, D. A., and Burchard, W., An analysis of the cooperative conformational transitions in the cellulose and amylose tricarbanilates, *Macromolecules* 15, 1350, 1982.
25. Perico, A., Mormino, M., Urbani, R., Cesàro, A., Tylianakis, E., Dais, P., and Brant, D. A., Local dynamics of carbohydrates: I. dynamics of simple glucans with different chain linkages, *J. Phys. Chem B*, 103, 8162, 1999.
26. Cowie, J. M. G., Some general features of T_g-M relations for oligomers and amorphous polymers, *Eur. Polym. J.*, 11, 297, 1975.
27. Levine, H. and Slade, L., A polymer physiochemical approach to the study of commercial starch hydrolysis products (SHPs), *Carbohydr. Polym.*, 6, 213, 1986.
28. Slade, L. and Levine, H., Non-equilibrium behavior of small carbohydrate-water systems, *Pure Appl. Chem.*, 60, 1841, 1988.
29. Orford, P. D., Parker, R., Ring, S. G., and Smith, A. C., Effect of water as a diluent on the glass transition behaviour of malto-oligosaccharides, amylose and amylopectin, *Int. J. Biol. Macromol.*, 11, 96, 1989.
30. Scandola, M., Ceccorulli, G., and Doi, Y., Viscoelastic relaxations and thermal properties of bacterial poly(3-hydroxybutyrate-CO-3-hydroxyvalerate) and poly(3-hydroxybutyrate-(0-4-hydroxybutyrate), *Int. J. Biol. Macromol.*, 12, 112, 1990.
31. Fox, T. G., Influence of diluent and of copolymer composition on the glass temperature of a polymer system, *Bull. Am. Phys. Soc.*, 1, 123, 1956.
32. Heijboer, J., Molecular origin of relaxation in polymers, *Ann. N.Y. Acad. Sci.*, 279, 104, 1976.
33. Kolarik, J., Secondary relaxations in glassy polymers: hydrophilic polymethacrylates and polyacrylates, *Adv. Polym. Sci.*, 46, 119, 1986.
34. Reiner, M. R., The Deborah number, *Phys. Today*, 17, 62, 1964.
35. Mc Crum, N. G., Read, B. E., and Williams, G., *Anelastic and Dielectric Effects in Polymeric Solids*, John Wiley & Sons, New York, 1967.
36. Struik, L. C. E., *Internal Stresses, Dimensional Stabilities and Molecular Orientations in Plastics*, John Wiley & Sons, Chichester, 1990, 48.

37. Flory, P. J., *Principles of Polymer Chemistry*, Cornell University Press, Ithaca, 1953.
38. Petri, H.-M., Schuld, N., and Wolf, B. A., Hitherto ignored effects of chain length on the Flory-Huggins interaction parameters in concentrated polymer solutions, *Macromolecules*, 28, 4975, 1995.
39. Roos, Y. H., *Phase Transitions in Foods*, Academic Press, San Diego, 1995.
40. Gordon, M. and Taylor, J. S., Ideal copolymers and the second-order transitions of synthetic rubbers. I. non crystalline co-polymers, *J. Appl. Chem.*, 2, 493, 1952.
41. Couchmann P. R. and Karasz, F. E., A classical thermodynamic discussion of the effect of composition on glass-transition temperatures, *Macromolecules*, 11, 117, 1978.
42. Couchman P. R., Compositional variation of glass-transition temperatures. 2. application of the thermodynamic theory to compatible polymer blends, *Macromolecules*, 11, 1156, 1978.
43. Blond, G., Simatos, D., Catte, M., and Dussap, C. G., Modeling of the water-sucrose state diagram below 0°C, *Carbohydr. Res.*, 298, 139, 1997.
44. Izzard, M. J., Ablett, S., and Lillford, P. J., Calorimetric study of the glass transition occurring in sucrose solutions, in *Food Polymers, Gels and Colloids*, Dickinson, E., Ed., The Royal Society of Chemistry, Cambridge, 1991, chap. 23.
45. Urbani, R., Sussich, F., Prejac, S., and Cesàro, A., Enthalpy relaxation and glass transition behaviour of sucrose by static and dynamic DSC, *Thermochim. Acta*, 304, 359, 1997.
46. ten Brinke, G., Karasz, F. E., and Ellis, T. S., Depression of glass transition temperatures of polymer networks by diluents, *Macromolecules*, 16, 244, 1983.
47. Pizzoli, M. and Scandola, M., Plasticizers, in *Polymeric Materials Encyclopedia*, Vol. 7, Salamone, J. C., Ed., CRC Press, Boca Raton, 1996, 5301.
48. Berghmans, H., Thermal transitions and gelation in polymer solutions, in *Calorimetry and Thermal Analysis of Polymers*, Mathot, V. B. F., Ed., Hanser Publ. Munich, 1994, chap. 8.
49. Nakamura, S. and Tobolsky, A. V., Viscoelastic properties of plasticized amylose films. *J. Appl. Polym. Sci.*, 11, 1371, 1967.
50. van den Berg, C., Vapour Sorption Equilibria and Other Water Starch Interactions: A Physiochemical Approach, Doctoral thesis, Agricultural University, Wageningen, Netherlands, 1981.
51. Zeleznak, K. J. and Hoseney, R. C., The glass transition of starch. *Cereal Chem.*, 64, 121, 1987.
52. Whittam, M. A., Noel, T. R., and Ring, S. G., Melting and glass/rubber transitions of starch polysaccharides, in *Food Polymers, Gels and Colloids*, Dickinson, E., Ed., The Royal Society of Chemistry, Cambridge, 1991, chap. 22.
53. Roos, Y. H. and Karel, M., Plasticizing effect of water on thermal behavior and crystallization of amorphous food models, *J. Food Sci.*, 56, 38, 1991.
54. Bizot, H., Le Bail, P., Leroux, B., Davy, J., Roger, P., and Buleon, A., Calorimetric evaluation of the glass transition in hydrated, linear and branched polyanhydroglucose compounds, *Carbohydr. Polym.*, 32, 33, 1997.
55. Donovan, J.W., Phase transitions of the starch-water system, *Biopolymers*, 18, 263, 1979.
56. Blanshard, J. M. V. and Lillford, P. J., Eds., *The Glassy State in Foods*, Nottingham University Press, Nottingham, 1993.
57. Shogren, R. L., Effect of moisture content on the melting and subsequent physical ageing of cornstarch, *Carbohydr. Polymers*, 19, 83, 1992.
58. Thiewes, H. J. and Steeneken, P. A. M., The glass transition and the sub-T_g endotherm of amorphous and native potato starch at low moisture content, *Carbohydr. Polymers*, 32, 123, 1997.

59. Cantor, C. R. and Schimmel, P. R., *Biophysical Chemistry. Part III. The Behavior of Biological Macromolecules,* Freeman and Co., San Francisco 1980.
60. Matveev, Y. I., Grinberg, V. Y., Sochava, I. V., and Tolstoguzov, V. B., Glass transition temperature of proteins. Calculation based on the additive contribution method and experimental data, *Food Hydrocolloids*, 11, 125, 1997.
61. Kalichevsky, M. T., Jaroszkiewicz, E. M., and Blanshard, J. M., A study of the glass transition of amylopectin-sugar mixtures, *Polymer*, 34, 346, 1993.
62. Kalichevsky, M. T., Jaroszkiewicz, E. M., and Blanshard, J. M., Glass transition of gluten. 1: gluten and gluten-sugar mixtures, *Int. J. Biol. Macromol.*, 14, 257, 1992.
63. Kalichevsky, M. T., Jaroszkiewicz, E. M., and Blanshard, J. M., Glass transition of gluten. 2: The effect of lipids and emulsifiers, *Int. J. Biol. Macromol.*, 14, 267, 1992.
64. Larsson, H. and Eliasson, A.-C., Phase separation of wheat flour dough studied by ultracentrifugation and stress relaxation. 1. influence of water content, *Cereal Chem.*, 73, 18, 1996.
65. Scandola, M., Ceccorulli, G., Pizzoli, M., and Pezzin, G., Further evidence of an unusual T_g-concentration dependence for plasticize polyvinylchloride, *Polym. Bull.*, 6, 653, 1982.
66. Kalichevsky, M. T. and Blanshard, J. M. V., A study of the effect of water on the glass transition of 1:1 mixtures of amylopectin, casein and gluten using DSC and DMTA, *Carbohydr. Polymers,* 19, 271, 1992.
67. Jouppila, K., Kansikas, J., and Roos, Y. H., Factors affecting crystallization kinetics in amorphous corn starch, *Carbohydr. Polymers,* 36, 143, 1998.
68. Lourdin, D., Coignard, L., Bizot, H., and Colonna, P., Influence of equilibrium relative humidity and plasticizer concentration on the water content and glass transition of starch materials, *Polymer*, 38, 5401, 1997.
69. Cherian, G. and Chinachoti, P., H-2 and O-17 nuclear magnetic resonance study of water in gluten in the glassy and rubbery state, *Cereal Chem.*, 73, 618, 1996.
70. Le Meste, M., Huang, V. T., Panama, J., Anderson, G., and Lentz, R., Glass transition of bread, *Cereal Foods World*, 37, 264, 1992.
71. Gontard, N. and Ring, S., Edible wheat gluten film: influence of water content on glass transition temperature, *J. Agric. Food Chem.*, 44, 3474, 1996.
72. Banks, W. and Greenwood, C. T., *Starch and its Components*, Edinburgh University Press, Edinburgh, 1975.
73. Clark, A. H. and Ross-Murphy, S. B., Structural and mechanical properties of biopolymer gels, *Adv. Polymer Sci.,* 83, 57, 1987.
74. Flory, P. J., Gels and gelling processes, *Faraday Discuss. Chem. Soc.*, 57, 7, 1974.
75. Roudaut, G., Maglione, M., van Dusschoten, D., and Le Meste, M., Molecular mobility in glass bread: a multispectroscopy approach, *Cereal Chem.*, 76, 70, 1999.
76. Roudaut, G., Maglione, M., and Le Meste, M., Relaxation below glass transition temperature in bread and its components, *Cereal Chem.*, 76, 78, 1999.
77. Waigh, T. A., Hopkinson, I., Donald, A.M., Butler, M.F., Heidelbach, F., and Riekel, C., Analysis of the native structure of starch granules with X-ray microfocus diffraction, *Macromolecules*, 30, 3813, 1997.
78. Gallant, D. J., Bouchet, B., and Baldwin, P. M., Microscopy of starch: evidence of a new level of granule organization, *Carbohydr. Polym.*, 32, 177, 1997.
79. Calvert, P., The structure of starch, *Nature*, 398, 338, 1997.
80. Hebeda, R. E. and Zobel, H. F., Eds., *Baked Goods Freshness*, Marcel Dekker, New York, 1996.
81. Slade, L. and Levine, H., Water and the glass transition. *J. Food Eng.*, 24, 431, 1995.
82. Morawetz, H., *Polymers — The Origin and Growth of Science*, John Wiley & Sons, New York, 1985, Part I.

3 Macromolecular Aspects of Bread Staling

Roger Parker and Stephen G. Ring

CONTENTS

INTRODUCTION

Bread staling is a general term that describes a time-dependent loss in quality of flavor and texture. The latter aspect is the focus of this article. From a molecular perspective, the main structural components of a processed cereal product such as bread are macromolecular, including starch, cell wall polysaccharides, and cereal proteins. The contribution of these components to the mechanical properties of the product may be modified by the presence of low molecular weight species, the most important of which is probably water but also includes salts, low molecular weight carbohydrates, and lipids. The molecular changes which occur during staling have been characterized and reviewed.[1] The baked product consists of a complex, multiphase biopolymer mixture and various low molecular weight species. The mechanical properties of such materials will be influenced by the phase behavior of the mixture and the dynamics of the molecules in each phase. It is this underlying physical chemistry of cereal biopolymers which we wish to examine to gain insight into the staling phenomenon. General principles are described for the behavior of related model systems, and their relevance to the bread staling problem is discussed. To develop this theme we first need to consider some molecular characteristics of the main biopolymers, starch and gluten.

0-8493-8790-6/01/$0.00+$.50

THE MACROMOLECULAR COMPONENTS

Starch consists of two main polysaccharides, one of which, amylose, is essentially linear; the other, amylopectin, is highly branched. Although fine structure is dependent on botanical origin,[2,3] for many amylopectins there is an essentially bimodal distribution of constituent chains. Current models of amylopectin structure[4] depict short linear chains, 10 to 20 units long, arranged in clusters on longer chains. Starch occurs naturally as partially crystalline water-insoluble granules. A number of crystalline forms are known. The A form,[5] which is found in most cereal starches, including wheat, consists of starch double helices packed into a monoclinic array. The B form,[6] which is found in some tuber starches and high amylose cereal starches, is a more highly hydrated and open structure consisting of double helices in a hexagonal array. In the starch granule, the amylopectin component forms the crystalline domains with a relatively short length of chain, at 10 to 15 units long.[7-9]

Gluten proteins are the major storage proteins of wheat, and consist of a complex mixture. The detailed structure of these polymers has been extensively researched and reviewed.[10] They can be broadly classified on the basis of their solubility in aqueous solvents. The gliadins are soluble in aqueous alcohols, while the glutenins are insoluble. A possible molecular origin of this insolubility is that glutenins form network structures crosslinked by disulphide bonds involving cysteine residues. Reduction of this linkage converts the polymers to alcohol-soluble monomers. The high molecular weight glutenin fraction is thought to contribute most to the elastic behavior of the gluten mixture. To gain insight into the viscoelastic behavior of glutenin, a comparison with the animal protein elastin is often made. Repeating peptide sequences are found in elastin which form β-spiral structures, and an intuitive view is that these structures behave as molecular springs. High molecular weight glutenins have comparable repeat motifs which form the same type of spiral structures. While on a molecular scale this notion of molecular springiness is attractive, the elastic properties of the material will also reflect longer distance interactions such as crosslinking.

MELTING, DISSOLUTION, AND GELATINIZATION

Starch is usually processed by heating in the presence of water which disrupts the native crystalline structure, a phenomenon known as gelatinization. In an excess of water (>90%w/w) above a characteristic temperature known as the gelatinization temperature, the starch granule loses its crystalline order and swells irreversibly to many times its original size. At the same time, the starch polysaccharide amylose is preferentially solubilized. At temperatures of less than 100°C and in the absence of mechanical shear, the swollen granules, enriched in amylopectin, maintain their integrity. The increase in volume fraction of starch granules in the suspension leads to an increase in viscosity of the paste. At concentrations greater than approximately 6%w/w, the gelatinized granules fill the available volume, producing a viscoelastic material. The properties of this material are influenced by the deformability of the swollen starch granule[11] and hence the amylopectin concentration in the granule.

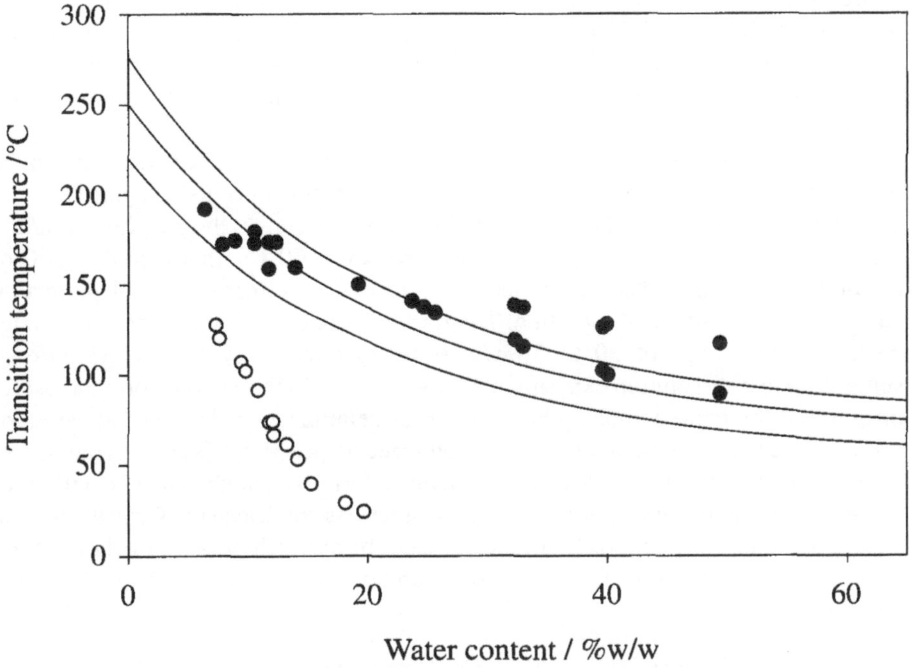

FIGURE 1 The effect of water content on the melting and glass transitions of starch. Key: solid line, estimates melting temperature of A-type short-chain amylose crystals with degree of polymerization (DP) 10, 12, 14;[18,19] ○, glass transition of pregelatinized starch;[34] ●, melting transition of granular wheat starch.[20]

One of the essential elements of the gelatinization process is the loss of crystalline order.[12] There have been a number of studies on the gelatinization process.[13] One of the difficulties in comparing this process to the melting and dissolution of synthetic polymer crystals is that the starch granule is a complex, partially crystalline structure, and gelatinization may be influenced by a number of factors. To gain deeper insight into the role of crystal melting and dissolution in gelatinization, it is necessary to examine highly crystalline materials, preferably single crystals of macroscopic size.

Short linear chains of amylose crystallize readily, and under appropriate conditions, highly crystalline materials of the different polymorphic forms of starch are obtained.[14-17] The effect of chain length on the process can be examined, with the expectation that the longer the length of chain involved in the crystalline unit, the higher the melting temperature — an expectation which has been confirmed experimentally.[18] A schematic diagram of the expected composition dependence of the melting of A type crystallites on water content is shown in Figure 1. The melting temperature of the low water-content material is experimentally inaccessible due to thermal degradation, polymorphic changes, and amorphisation on drying. In common with polymeric crystalline materials, the addition of a diluent, in this case water, depresses the observed melting temperature, T_m. The expected melting temperature at high water contents for an A type starch crystallite formed from chains 12 units in length is about 75°C. In comparison, the predicted high molecular weight limit of T_m in excess water is 150°C. With decreasing water content, the melting temperature rises, reaching a predicted 150°C for chains that are 12 units in length, at a water content of 16%w/w.

The crystalline form of starch also affects the dissolution behavior. At a fixed water content, the melting and dissolution of the B form occurs at approximately 20°C lower temperature.[19] For the crystallites in wheat starch, the gelatinization/dissolution temperature in excess water is approximately 60°C rather than the 75°C of the highly crystalline material, probably indicating an effect of crystal perfection and size on the observed dissolution. The form of the increase with decreasing water content in the range of 50 to 5%w/w water is remarkably similar[20] (Figure 1), suggesting that dissolution experiments on very crystalline materials provide useful insight into the more complex phenomenon of gelatinization. Further indication of use is gained from experiments on smooth-seeded pea starches,[21] the granules of which contain a mixture of A and B crystalline forms of starch. The dissolution of B domains at a lower temperature than A domains was confirmed by X-ray diffraction experiments on the gelatinization process. Current research on starch gelatinization examines the influence of the amorphous component of starch on the melting and dissolution process.

Interactions with Water and Low Molecular Weight Solutes

To understand the role of water as a diluent, it is useful to examine predictive relationships that describe the composition dependence of melting.[12,22] The classic description of the compositional dependence of polymer melting in the presence of a diluent is given by

$$1/T_m = 1/T_m^0 + (R/\Delta H_u) \cdot (V_u/V_1) \cdot [\nu_1 - \chi \nu_1^2] \quad (1)$$

where T_m^0 is the melting temperature of the pure polymer, V_u and V_1 are the molar volumes of polymer repeating unit and diluent, respectively, and v_1 is the diluent volume fraction. ΔH_u is the enthalpy of fusion per repeating unit and χ is the Flory-Huggins interaction parameter,[22,23] which characterizes the interaction energy per solvent molecule.

This relationship predicts that the smaller the diluent size relative to that of the polymer repeating unit (in this case an anhydroglucose unit), and the stronger the favorable interaction between the diluent and polymer, the greater the depression in T_m. It is well known that starch polysaccharides precipitate from aqueous solution at room temperature, indicating that water interacts weakly with the starch chain, and at room temperature it can be considered a poor solvent. Physico-chemical measurements of water sorption behavior provide a more quantitative estimate of solvent quality. The values of the parameter χ obtained for concentrated aqueous mixtures of maltoligomers were 0.7 to 0.8 at room temperature, confirming that water is a poor solvent under these conditions. To refine this approach, more information is needed on the temperature and composition dependence of χ. The relatively small molecular size of water, however, is predicted to lead to a strong depression in T_m. The Flory-Huggins approach fits the available experimental data for starch crystallite dissolution reasonably well. The prediction that replacement of water by other larger water-soluble solutes that interact weakly with the starch chain, such as low molecular weight carbohydrates, would lead to an elevation of T_m has also been confirmed experimentally.[24,25]

These aspects of biopolymer behavior appear quite general, although the underlying physical chemistry may be presented from different perspectives. For example in the examination of protein denaturation as a function of water content[26] showed that the denaturation temperature for ovalbumin increased from 77°C to 114°C as the water content was reduced from 40%w/w to 10%w/w. Similarly, the addition of low molecular-weight carbohydrates to proteins results in increase in protein stability and a higher denaturation temperature. The physico-chemical explanation of these effects is based on the observation that most proteins at room temperature in mixed carbohydrate/water solvents have a preferred interaction with water, i.e., carbohydrate is excluded from the domain of the protein.[27,28] The replacement of water by carbohydrate results in a decrease in solvent quality for the protein. A manifestation of this effect is an increase in the denaturation temperature. The gluten proteins should show the same type of behavior in that they will be preferentially hydrated in the presence of low molecular weight carbohydrates, and elements of secondary structure would be stabilized.

BIOPOLYMER/BIOPOLYMER INTERACTIONS

A further aspect of phase behavior that should be considered is immiscibility of biopolymer mixtures. Immiscibility of concentrated solutions of chemically different synthetic polymers is a well-described phenomena and can result from either an unfavorable energetic interaction between the polymers, or from the solvent interacting differently with the polymers.[29] Biopolymers show the same general type of behavior. Even two chemically similar biopolymers, amylose and amylopectin, differing primarily in their extent of branching, phase separate from concentrated solution.[30] This is an example of a small difference in molecular structure producing a potentially large effect on the microstructure of mixed biopolymeric material. In the case of cereal products, which have a much more complex bioploymer composition, the underlying phase structure is likely to be similarly complex. Additionally, after processing only some of the biopolymers are truly solubilized, and in reality a heterogeneous structure is present. For example, immediately after heat processing of starch in breadmaking, swollen gelatinized granules and partially solubilized amylose will be present rather than a homogeneous amylose/amylopectin/water mixture. Nevertheless, any solvent will distribute itself between phases depending on its relative affinity for water. For example, if the same amounts of two glucan polymers, amylose and dextran, are mixed in concentrated solution[31] at a temperature where crystallization is not observed on practical time scales, phase separation is observed with the formation of dextran-rich and amylose-rich phases. The phase volume of the dextran phase is much larger, i.e., water has a preferred interaction with the dextran. The way that water partitions between different phases can therefore have a potentially marked effect on microstructure and phase volumes. Stability can also be affected. For example, consider a concentrated solution containing 40% w/w water with 30% w/w each of dextran and amylopectin at room temperature. Storage at room temperature for this composition with even distribution of water would represent a quench of 80°C (T_m ~ 100°C) for chains 12 units in length, and the amylopectin in the amylopectin-rich phase would tend to crystallize. If the amylopectin phase was phase-concentrated by the presence of dextran, the resulting increase in amylopectin concentration would increase the driving force for crystallization.

Molecular mobility, which will affect the observed rate of crystallization and the mechanical properties should also be examined.

GLASS TRANSITION BEHAVIOR

Included in the schematic of Figure 1 is another transition which has an important influence on material properties[32] — the glass transition, which is characterized by the glass transition temperature T_g. Crystalline anhydrous β-D-glucose melts at 150°C. If this melt is cooled at a more rapid rate than the rate of crystallization, viscosity of the liquid will progressively increase until at 7°C viscosity reaches about 10^{12} Pa s.[33] At these enormous viscosities, the material behaves as a brittle solid. Glasses may be stable to crystallization for many years, the enormous viscosity of the glass arresting crystal nucleation and growth. At the calorimetric T_g, a sharp change in heat capacity is observed, indicative of a change from solid-like to liquid-like behavior within the timescale of the calorimetric experiment. This provides an experimentally convenient method for the determination of T_g as the midpoint of the step change. The calorimetric approach is useful for determination of the glass transition behavior of simple amorphous biopolymer mixtures. The glass transition line in Figure 1 shows the composition dependence of the T_g of a starch/water mixture.[34,35] The T_g of the dry polymer is experimentally inaccessible because thermal degradation intervenes. The addition of water has a strong plasticizing effect, causing a marked depression in T_g until at 20%w/w water, T_g reaches room temperature. Extrapolation of the available T_g data on starch/water mixtures to dryness results in an estimate of 225°C for the dry polymer. Wheat gluten proteins show broadly similar behavior,[36,37] where the extrapolated T_gs of dry cereal proteins fall in the range of 124 to 145°C. Addition of 15%w/w water depresses the glass transition to room temperature or slightly below. The isolated gluten proteins show a marked calorimetric glass transition the same way that the animal protein elastin does. This indicates that even though regions of the protein chain may have a preferred local conformation, at temperatures above T_g, the protein chain is sufficiently mobile to behave the same as amorphous synthetic polymers.

At low water contents, the T_g of the biopolymer/water mixture is very sensitive to small changes in water content, and by implication some of the material properties are equally sensitive. What is the situation, then, at higher water contents such as those found in bread? At a water content of 40% w/w, the extrapolated T_g of an amylopectin/water mixture is –58°C, and that of a HMW glutenin/water mixture –75°C. If there is an even distribution of water in the biopolymeric matrix of bread, its T_g is well below room temperature. The water content varies within the bread and may range from 25 to 30% at the crust to 45% in the middle of the loaf. Although the glass transition appears to be relatively remote, it is worth posing the question — is the transition sufficiently close to exert some effect on material properties? To answer this question, information is needed on the temperature dependence of mechanical properties for these amorphous materials.[38]

Many synthetic amorphous polymers undergo an enormous change in viscoelastic properties in the vicinity of the glass transition. The behavior of many amorphous biopolymeric materials is expected to be similar. An expression which has been shown to be widely applicable in describing this temperature dependence is,[38]

$$\log a_T = -c_{1,0}(T - T_0)/(c_{2,0} + T - T_0) \tag{2}$$

where a_T is the ratio of relaxation times at the temperature, T, and a reference temperature T_0. If T_g becomes the reference temperature then,

$$\log a_T = -c_{1g}(T - T_g)/(c_{2g} + T - T_g) \tag{3}$$

with values of the coefficients, c_{1g} and c_{2g}, obtained from fitting data on a range of synthetic polymers, 17.44 K^{-1} and 51.6 K, respectively. At the calorimetric glass transition, the shear viscosity, η, is of the order of 10^{12} Pa s and the shear stress relaxation time, t, is of the order of 100s and is given by $t = h/G_\infty$, where G_∞ is the high frequency limit of the shear modulus, which is 10^{10} P_a for many materials. The temperature and composition dependence of relative relaxation time, a_T, is shown schematically in Figure 2 for an amorphous, high molecular weight biopolymer/water mixture, and is based on the observed glass transition behavior of starch/water mixtures. As the glass transition is approached, either through reducing temperature or water content, a marked change in relaxation behavior is predicted. Structural relaxation and molecular mobility slows and, as a result, there is a marked change in material properties for a relatively small change in water content or temperature. Although the effect is most marked as the glass transition is approached, even for the average water contents found in bread, at room temperature there is a marked change in relaxation behavior (i.e., orders of magnitude) for relatively small changes in water content. This change in molecular mobility will have an impact on both stability to crystallization and local viscosity.

The above discussion was concerned with a hypothetical biopolymer/water mixture. Experimental data on the composition and temperature dependence of the mechanical behavior of cereal biopolymers is needed. For a more detailed discussion on the use of Equation 3 to describe the behavior of synthetic polymers, the reader is referred to the work of Ferry.[38] The use of this approach to describe the mechanical behavior of more complex systems or temperature dependence of other properties, while potentially useful, should be treated with caution.

At higher water contents such as those found in the middle of bread, the T_g is relatively remote.[39] The structural relaxation and polymer mobility is relatively rapid, and the observed viscoelastic behavior becomes strongly influenced by the extent of chain entanglement. The entanglements can behave as temporary crosslinks, increasing the observed elastic component of the viscoelastic material. At sufficiently

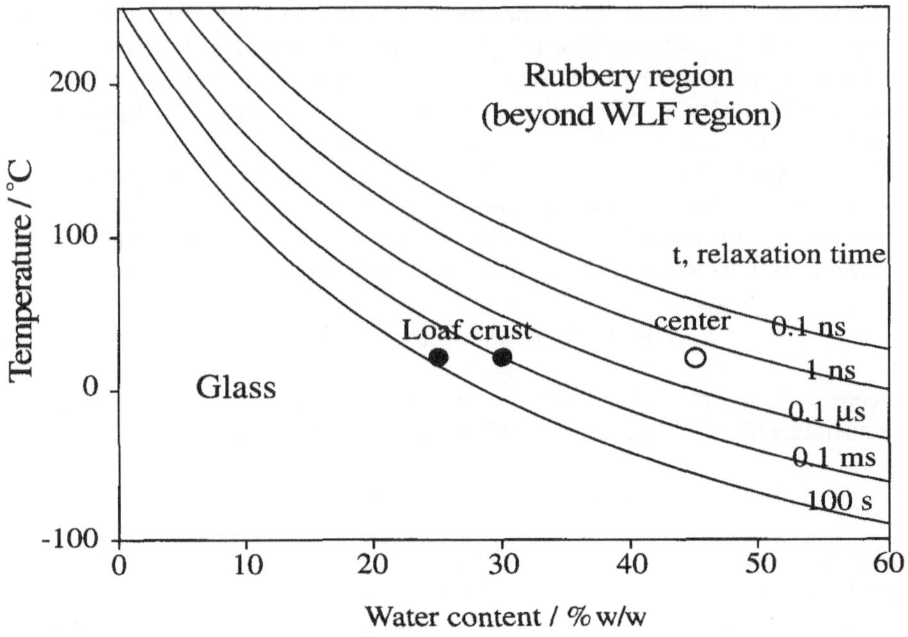

FIGURE 2 Effect of temperature and water content upon relaxation time calculated from the glass transition curve[34] and assuming WLF behavior.[38] Water content of loaf crust and center shown at 20°C.

long experimental times the material will still flow. For highly branched polymers, such as amylopectin, the presence of chain branches which hinder disentanglement and flow in concentrated systems is expected to have a strong effect. This behavior may be further modified through the introduction of crosslinks, which may be covalent or formed through secondary associations. This crosslinking results in the formation of a three-dimensional polymer network. If this network is sufficiently permanent, even at long experimental times the material properties will still exhibit elastic behavior. Crosslinking and network formation is worth examining in more detail as they can have a marked effect on material properties.

BIOPOLYMER AGING IN BREAD

If a gelatinized starch/water mixture is cooled to room temperature, there is a driving force favoring crystallization. The processes which occur on cooling are known as retrogradation.[40-44] The high molecular-weight linear polymer, amylose, has a strong driving force with an effective quench of at least 120°C if the material is held at room temperature. Such a strong quench is not conducive to the formation of very crystalline materials. If a concentrated aqueous amylose solution is cooled to room temperature, there is a very rapid precipitation/phase separation process. The clear amylose solution rapidly becomes opaque, indicating formation of polymer aggregates at least the order of the wavelength of light in size. For sufficiently concentrated solutions (typically greater than 1 to 2% w/w), network formation is observed through the formation of an elastic gel. Network formation, as assessed by the development of stiffness, is generally complete within a few hours at room temperature. Electron microscopic examination of the network reveals coarse network strands which consist of assemblies of many individual polymer strands.[45] Subsequent to this precipitation, a slow crystallization of the amylose is observed by X-ray diffraction over the course of a few days. This crystallization does not have a marked effect on material properties; these materials remain poorly crystalline at the end of this time.

The effective quench for formation of crystallites from short linear chains is much more modest. If concentrated amylopectin/water mixtures are quenched to room temperature or below, a slow crystallization of the amylopectin is observed as assessed by X-ray diffraction. In this case crystallization is associated with the development of stiffness of the material, and it typically takes days or weeks to approach a plateau value. At the end of this time, the extent of crystallinity of the amylopectin is comparable to that found in the native starch granule, i.e., about 30%. The length of chain involved in the crystalline unit is relatively short. Examination of the behavior of amylopectins from different botanical sources[46,47] showed that the longer the length and abundance of the short chain fraction of amylopectin, the greater the tendency to retrograde and crystallize from aqueous solution. Wheat amylopectin, whose short chain fraction is relatively short, generally shows a reduced tendency to retrograde. Even so, it can lead to significant time-dependent changes in material properties.

For materials prepared by gelatinizing starch and then cooling to room temperature, the precipitation of amylose is a very rapid process. The crystallization of amylopectin leads to an increase in stiffness of the gelatinized granule and a slow firming of the material. Amylopectin crystallization and the associated firming can be abolished by reheating to 60°C. In bread, this crystallization is recognized as making a significant contribution to the staling process.

Structural Relaxation

For completeness sake, it is worth briefly examining the phenomenon of structural relaxation in glassy materials. Although not directly relevant to most bread products, it is potentially relevant to the aging behavior of low-moisture cereal products which are glassy. The equilibrium structure of a glass is temperature dependent. For example, it is expected that reducing temperature would increase the density of the material. In Figure 2 the relative relaxation time is shown to increase as the glass transition is approached through lowering temperature or water content. A typical structural relaxation time at T_g is in the 100's. On further cooling, this relaxation time will progressively increase. If an amorphous material is rapidly quenched far into the glass, the structural relaxation time may be so high that the amorphous structure, and resulting density, is effectively frozen in. At very long times it will have a fully relaxed, equilibrium structure. If it is cooled, further structural relaxations and rearrangements within the undercooled liquid will occur until, given sufficient time, a new equilibrium structure is obtained. The structure of the undercooled amorphous polymer liquid can therefore show a dependence on time and thermal history. This relaxation behavior has been extensively studied, particularly in polymers and inorganic glasses. The temperature-dependent changes in bonding between molecules and their configuration are associated with changes in volume, enthalpy, heat capacity, and material properties, including mechanical behavior and diffusivity. The observed changes in mechanical behavior of amorphous glasses with time is often described as embrittlement. This phenomenon should be particularly relevant to the aging of low water-content cereal products. Relationships have been developed for synthetic materials which describe the process well and are of practical utility in predicting the aging as a function of thermal history.[48]

ENZYME EFFECTS

It is widely reported that amylolytic enzymes can reduce the rate of staling of bread, although the precise mechanism through which they have their effect is less clear.[49] Experiments on amylopectin solutions show that addition of the exo-acting enzyme β-amylase inhibited the retrogradation of amylopectin.[50] This effect was concluded to be a result of the shortening of external amylopectin chains to a length which would not crystallize from aqueous solution at room temperature. The role of α-amylases in delaying retrogradation is less obvious. The patent literature refers to the use of low doses of bacterial or fungal α-amylase whose activity is reasonably thermally stable. While the action on native starch granules might be low, the presence of amylolytic activity after starch gelatinization is expected to modify the mechanical behavior of the starch-rich phase. A limited amylolysis can be effective in inhibiting retrogradation, while not producing undesirable gumminess in the product. While this amylolysis might be expected to soften the texture of starch-rich materials through limited depolymerization of solubilized amylose and amylopectin chains, it is less obvious how a limited amylolysis would affect the crystallization of amylopectin chains in gelatinized starch granules. The formation of low molecular weight species would increase the affinity of the starch-rich phase for water, thereby reducing starch concentration and the driving force for the crystallization of starch chains. A change in water content of the starch-rich phase would also modify its mechanical characteristics. It may also be proposed that short maltodextrin chains modify retrogradation behavior by interfering with the crystallization of amylopectin chains.[51] These explanations are speculative and further research is needed to elucidate the precise mechanism of action.

EFFECT OF SURFACTANTS

If amylose in hot aqueous solution is cooled to room temperature, the amylose precipitates and retrogrades with formation of a partially crystalline precipitate or gel. The precipitate needs to be heated to >120°C to initiate dissolution. If the aqueous solution is saturated with 1-butanol prior to cooling, a precipitate is still formed which will dissolve on heating to 100°C. This precipitate consists of amylose in a single helical form and can contain the 1-butanol as a complexed guest molecule.

On the basis of X-ray diffraction experiments, the complex may be either crystalline or amorphous.[52] The amorphous material is formed on rapid cooling, while slow cooling promotes the formation of crystalline material. The addition of a hydrophobic species prevents the formation of the B-crystalline form. The single helical complex is the preferred product.[53-55] Depending on the hydrophobic species, a conformational change could be induced to a helical form which does not readily crystallize or precipitate. The addition of hydrophobic surfactants to starch-rich products would modify the retrogradation behavior by favoring the formation of single helical conformations. Depending on the conditions, this could be single helices in solution or an amorphous/crystalline complex. While the interaction with amylose is relatively strong, favoring the formation of crystalline materials, the interaction with the shorter chains of amylopectin would be expected to be weaker. After addition of surfactants to bread products, the crystalline single helical V_h form of amylose is observed and the development of the B-form on aging of the bread is retarded.[1,56] The consequent effect on mechanical behavior of these conformational changes should depend on whether the surfactant favors the formation of solid or soluble complexes. In these complex multiphase/multicomponent systems, surfactants are likely to produce a diversity of effects.

REFERENCES

1. Zobel, H. F. and Kulp, K., The staling mechanism, in *Baked Goods Freshness: Technology, Evaluation and Inhibition of Staling,* Hebeda, R. E. and Zobel, H. F., Eds., Marcel Dekker, New York, 1996, chap. 1.
2. Hizukuri, S., Polymodal distribution of the chain lengths of amylopectins, and its significance, *Carbohydr. Res.,* 147, 342, 1986.
3. Hizukuri, S. and Maehara, Y., Fine-structure of wheat amylopectin — the mode of a-chain to b-chain binding, *Carbohydr. Res.,* 206, 145, 1990.
4. Manners, D. J., Recent developments in our understanding of amylopectin structure, *Carbohydr. Polym.,* 11, 87, 1989.
5. Imberty, A., Chanzy, H., Perez, S., Buleon, A., and Tran, V., The double-helical nature of the crystalline part of A-starch, *J. Mol. Biol.,* 201, 365, 1988.
6. Imberty, A. and Perez, S., A revisit to the 3-dimensional structure of B-type starch, *Biopolymers,* 27, 1205, 1988.
7. Banks, W. and Greenwood, C. T., *Starch and its Components*, Edinburgh University Press, Edinburgh, 1975.
8. French, D., Organization of starch granules, in *Starch Chemistry and Technology,* Whistler, R. L., Paschall, E. F., and BeMiller, J. N., Eds., Academic Press, New York, 1984.
9. Guilbot, A. and Mercier, C., Starch, in *The Polysaccharides,* Vol. 3, Aspinall, G. O., Ed., Academic Press, New York, 1985.
10. Shewry, P. R., Tatham, A. S., Barro, F., Barcelo, P., and Lazzeri, P., Biotechnolgy of breadmaking: unraveling and manipulating the multi-protein gluten complex, *Biotechnology,* 13, 1185, 1995.
11. Ellis, H. S., Ring, S. G., and Whittam, M. A., A comparison of the viscous behaviour of wheat and maize starch pastes, *J. Cereal Sci.,* 10, 33, 1989.
12. Donovan, J. W., Phase transitions of the starch-water system, *Biopolymers*, 18, 263, 1979.

13. Cooke, D. and Gidley, M. J., Loss of crystalline and molecular order during starch gelatinization — origin of the enthalpic transition. *Carbohydr. Res.*, 227, 103, 1992.
14. Pfannemuller, B., Influence of chain-length of short monodispersed amyloses on the formation of A-type and B-type X-ray diffraction patterns, *Int. J. Biol. Macromol.*, 9, 105, 1987.
15. Ring, S. G., Colonna, P., Miles, M. J., Morris, V.J., and Turner, R. J., Spherulitic crystallization of short chain amylose, *Int. J. Biol. Macromol.*, 9, 158, 1987.
16. Gidley, M. J. and Bulpin, P. V., Crystallization of malto-oligosaccharides as models of the crystalline forms of starch — minimum chain-length requirement for the formation of double helices, *Carbohydr. Res.*, 161, 291, 1987.
17. Helbert, W., Chanzy, H., Planchot, V., Buleon, A., and Colonna, P., Morphological and structural features of amylose spherocrystals of A-type, *Int. J. Biol. Macromol.*, 15, 183, 1993.
18. Moates, G. K., Noel, T. R., Parker, R., and Ring, S. G., The effect of chain length on the dissolution of the B-crystalline polymorph of starch, *Carbohydr. Res.*, 298, 327, 1997.
19. Whittam, M. A., Noel, T. R., and Ring, S. G., Melting behavior of A-type and B-type crystalline starch, *Int. J. Biol. Macromol.*, 12, 359, 1990.
20. Burt, D. J. and Russell, P. L., Gelatinisation of low water-content wheat starch water mixtures — a combined study by differential scanning calorimetry and light microscopy, *Starch*, 35, 354, 1983.
21. Bogracheva, T. Y., Morris, V. J., Ring, S. G., and Hedley, C. L., The granular structure of c-type pea starch and its role in gelatinization, *Biopolymers*, 45, 323, 1988.
22. Flory, P. J., *Principles of Polymer Chemistry*, 1st ed., Cornell University Press, Cornell, 1953.
23. Orwoll, R. A., The polymer-solvent interaction parameter, χ. *Rubber Chem. Technol.*, 50, 452, 1977.
24. Lelievre, J., Theory of gelatinization in a starch-water-solute system, *Polymer*, 17, 854, 1976.
25. Moates, G. K., Parker, R., and Ring, S. G., Preferential solvent interactions and the dissolution of the B-type crystalline polymorph of starch in aqueous solutions, *Carbohydr. Res.*, 313, 225, 1998.
26. Rupley, J. A. and Careri, G., Protein hydration and function, *Adv. Protein Chem.*, 41, 37, 1991.
27. Lee, J. C., Gekko, K., and Timasheff, S. N., Measurements of preferential solvent interactions by densimetric techniques, *Methods in Enzymol.*, 61, 26, 1979.
28. Arakawa, T. and Timasheff, S. N., Stabilization of protein structure by sugars, *Biochemistry*, 21, 6536, 1982.
29. Altena, F. W. and Smolders, C. A., Calculation of liquid-liquid phase separation in a ternary system of a polymer in a mixture of a solvent and a non-solvent, *Macromolecules*, 15, 1491, 1982.
30. Kalichevsky, M. T. and Ring, S. G., Incompatibility of amylose and amylopectin in aqueous-solution, *Carbohydr. Res.*, 162, 323, 1987.
31. Kalichevsky, M. T., Orford, P. D., and Ring, S. G., The incompatibility of concentrated aqueous-solutions of dextran and amylose and its effect on amylose gelation, *Carbohydr. Polym.* 6, 145, 1986.
32. Slade, L. and Levine, H., Glass transitions and water-food structure interactions, *Adv. Food Nutr. Res.*, 38, 103, 1995.
33. Parks, G. S., Barton, L. E., Spaght, M. E., and Richardson, J.W., The viscosity of undercooled liquid glucose, *Physics*, 8, 193, 1934.

34. Zeleznak, K. J. and Hoseney, R. C., The glass transition in starch, *Cereal Chem.*, 64, 121, 1987.
35. Orford, P. D., Parker, R., Ring, S. G., and Smith, A. C., Effect of water as a diluent on the glass-transition behavior of malto-oligosaccharides, amylose and amylopectin, *Int. J. Biol. Macromol.*, 11, 91, 1989.
36. Kalichevsky, M. T., Jaroszkiewicz, E. M., and Blanshard, J. M. V., Glass transition of gluten. 2. The effect of lipids and emulsifiers, *Int. J. Biol. Macromol.*, 14, 267, 1992.
37. Noel, T. R., Parker, R., Ring, S. G., and Tatham, A. S., Glass transition behaviour of wheat gluten proteins, *Int. J. Biol. Macromol.*, 17, 81, 1995.
38. Ferry, J. D., *Viscoelastic properties of polymers*, John Wiley and Sons, New York, 1980.
39. Janssen, A. M., van Vliet, T., and Vereijken, J. M., Rheological behaviour of wheat glutens at small and large deformations. Comparison of two glutens differing in bread making potential, *J. Cereal Sci.*, 23, 19, 1996.
40. Miles, M. J., Morris, V. J., and Ring, S. G., Gelation of amylose, *Carbohydr. Res.*, 135, 257, 1985.
41. Miles, M.J., Morris, V.J., and Ring, S.G., Roles of amylose and amylopectin in the gelation and retrogradation of starch, *Carbohydr. Res.*, 135, 269, 1985.
42. Ring, S. G., Colonna, P., Ianson, K. J., Kalichevsky, M. T., Miles, M. J., Morris, V. J., and Orford, P. D., The gelation and crystallization of amylopectin, *Carbohydr. Res.*, 162, 277, 1987.
43. Gidley, M. J., Molecular mechanisms underlying amylose aggregation and gelation, *Macromolecules*, 22, 351, 1989.
44. Gidley, M. J. and Bulpin, P. V., Aggregation of amylose in aqueous systems — the effect of chain-length on phase-behavior and aggregation kinetics, *Macromolecules*, 22, 341, 1989.
45. Leloup, V. M., Colonna, P., Ring, S. G., Roberts, K., and Wells, B., Microstructure of amylose gels, *Carbohydr. Polym.*, 18, 189, 1992.
46. Kalichevsky, M. T., Orford, P. D., and Ring, S. G., The retrogradation and gelation of amylopectins from various botanical sources, *Carbohydr. Res.*, 198, 49, 1990.
47. Shi, Y. C. and Seib, P. A., The structure of 4 waxy starches related to gelatinization and retrogradation, *Carbohydr. Res.*, 227, 131, 1992.
48. Hodge, I. M., Enthalpy relaxation and recovery in amorphous materials, *J. Non-Crystal. Solids*, 169, 211, 1994.
49. Bowles, L. K., Amylolytic enzymes, in *Baked Goods Freshness: Technology, Evaluation and Inhbition Of Staling*, Hebeda, R. E. and Zobel, H. F., Eds., Marcel Dekker, New York, 1996, chap. 3.
50. Wursch, P. and Gumy, D., Inhibition of amylopectin retrogradation by partial beta-amylolysis, *Carbohydr. Res.*, 256, 129, 1994.
51. Defloor, I. and Delcour, J. A., Impact of maltodextrins and antistaling enzymes on the differential scanning calorimetry staling endotherm of baked bread doughs, *J. Agric. Food Chem.*, 47, 737, 1999.
52. Whittam, M. A., Orford, P. D., Ring, S. G., Clark, S. A., Parker, M. L., Cairns, P., and Miles, M. J., Aqueous dissolution of crystalline and amorphous amylose-alcohol complexes, *Int. J. Biol. Macromol.*, 11, 339, 1989.
53. Buleon, A., Delage, M. M., Brisson, J., and Chanzy, H., Single-crystals of V-amylose complexed with isopropanol and acetone, *Int. J. Biol. Macromol.*, 12, 25, 1990.
54. Godet, M. C., Tran, V., Delage, M. M., and Buleon, A., Molecular modeling of the specific interactions involved in the amylose complexation by fatty-acids, *Int. J. Biol. Macromol.*, 15, 11, 1993.

55. Godet, M. C., Tran, V., Colonna, P., Buleon, A., and Pezolet, M., Inclusion exclusion of fatty-acids in amylose complexes as a function of the fatty-acid chain-length, *Int. J. Biol. Macromol.*, 17, 405, 1995.
56. Knightly, W. H., Surfactants, in *Baked Goods Freshness: Technology, Evaluation and Inhbition of Staling,* Hebeda, R. E. and Zobel, H. F., Eds., Marcel Dekker, New York, 1996, chap. 2.

4 An Interpretation of the Rheological Behavior of Wheat Flour Dough Based on Fundamental Tests

Paolo Masi, Silvana Cavella, and Laura Piazza

CONTENTS

INTRODUCTION

The unique rheological response of wheat flour doughs contributes to their versatility from a technological point of view. Flour doughs can behave simultaneously as a viscous liquid and an elastic solid.[1] A viscous liquid flows by the gravitational force, where geometry deformation and stress relaxation are observed over time. Conversely, an elastic material recovers part of the deformation when the stress is removed, such as when forced to pass through a die, where the material expands at the die exit. Many foodstuffs and various engineering materials exhibit this behavior as a result of their complex structure.

The question of where the viscoelastic behavior of the wheat flour dough originated and how it affects the flour performance in baking operation, which is of paramount practical importance, remains a topic for further investigations. The

0-8493-8790-6/01/$0.00+$.50

complex roles played by various components of the flour and other ingredients in determining the quality and functionality of the finished product is not completely understood, despite ongoing research since the pioneering work of Schofield and Scott Blair.[2] This can partly be ascribed to the empirical research approach followed in the past, which resulted in experimental data that reflected the geometry and the dynamics of the test apparatus rather than the physical properties of the sample. Therefore, the results were helpful in describing the dough performance under specific process conditions but inappropriate for deriving structure-properties relationships. This chapter illustrates some recent data based on fundamental rheological tests which offer more basic viscoelastic information and related structural organization of the dough.

STRUCTURAL ORGANIZATION OF THE DOUGH

Flour dough is a complex mixture consisting of starch, (~80%), proteins, (~14%), lipids, (~4-5%), and pentosans, (~1–2%).[3] With adequate quantities of water, part of these flour components (i.e., pentosans, soluble proteins, or damaged starch) are dissolved into the liquid phase, while insoluble hydrated proteins and starch globules distribute uniformly throughout the entire volume, resulting in a sticky paste. To obtain a dough, flour and water must be mixed for an appropriate period of time. During mixing, shear and tensile stresses disperse the water molecules into the flour, promoting the structure of the dough.

The protein fraction (gluten) is the main contributor to dough structure formation and viscoelasticity. The viscoelastic character can be found in a lesser extent in rye and barley (which contain a smaller amount of gluten proteins), and is entirely absent from oat, maize, and sorghum (which are cereals completely lacking in gluten proteins).[4] Many models have been proposed to describe the molecular organization of proteins,[5,6] although none of them fully explains the viscoelastic behavior of dough under different circumstances.[7] To better illustrate the most reliable hypothesis proposed to depict the gluten functionality, an understanding of the molecular structure of the proteins and their behavior in the presence of water is necessary.

Gliadins, as well as globulins and albumins, are referred to as monomeric proteins, i.e., they are formed by a single sequence of amino acids which can link together through disulfide bonds via cysteine residues, and have molecular weight ranging from 30,000 to 80,000. Conversely, the molecular structure of the glutenins is much more complex due to the branched chain of sub-units held together by disulfide bonds (polymeric proteins). The chained sub-units consist of two types: those with low molecular weight in the range 40,000 to 55,000 (most abundant), and those with molecular weight ranging from 80,000 to 120,000.[8]

In a moist environment, many types of interactions can occur among portions of the same protein, between different proteins, and between proteins and flour components. These interactions are mostly temporary and can break and reform either when the dough is at rest or when it is deformed, reflecting the dual rheological character of dough. For example, among atoms of the same molecules, van der Waal interactions can be established whereby two portions of the same protein can link together by sharing a hydrogen atom, thus forming hydrogen bonds. Due to the thermodynamic incompatibility between the apolar protein side chains and the water, clusters of proteins may form in order to exclude water, thus originating hydrophobic links. From a rheological point of view, the most important type of interaction which can occur between and within proteins is the one which involves sulfhydryl groups. These groups are potentially capable of undergoing a disulfide-sulfhydryl interchange that involves the cleavage or reformation of disulfide bonds.

Upon addition of water, the soluble flour constituents dissociate and proteins hydrate. The resultant volume expansion leads to increased short and long range rotational mobility, promoting the formation of temporary junctions between proteins.[9] However, since the proteins are randomly distributed in the flour, weak secondary interactions are insufficient to hold together the constituents and generate a continuous network. During the mixing operation, the proteins experience tensile and shear stresses, thereby increasing their contact surface, which leads to additional interactions. Subsequently, contribution from the secondary forces becomes relevant, and proteins form a continuous, branched network. Due to the large molecular weight and complex conformation of the polymeric proteins, it has been suggested that in addition to these interactions, physical entanglements may be present in the gluten structure. This hypothesis, first presented by Ewart[10] then MacRitchie,[11] elucidated the strong similarity existing between the gluten structure of the dough and that of polymers whose viscoelastic behavior is often explained in terms of entangled macromolecules network. In this chapter, for rheological purposes, flour dough will be depicted as an homogeneous material formed by two continuous phases.[12] The first phase is the one formed by the water-swelled proteins, and the second phase consists of the water solution which uniformly surrounds this network in which starch granules are dispersed. Accordingly, the dough performance will be explained in terms of the configuration of the entangled protein network and its reciprocal interaction with the surrounding liquid phase. The configuration of the entangled gluten network will depend mainly on the composition of the proteins, their conformation, and their molecular weight distribution, while the interaction between the gluten network and the surrounding liquid phase will depend on flour composition.

EMPIRICAL TESTS

Until the late eighties, evaluation of the baking performance of flours of different varieties was commonly performed by empirical tests. These tests are based on the direct observation of the flour performance by subjecting a sample to complex actions which imitate the real environment in a baking stage, and estimating properties which are correlated to the breadmaking practice. The value of these estimates depends on the ability of the user to interpret the data relevant to specific baking formulas, procedure, and products. The farinograph, the alveograph, and the extensiograph, which are popular among bakery scientists and technologists have been developed for this purpose.[13]

The farinograph is commonly used to estimate the optimal amount of water which should be added to a particular flour, the dough development time, and dough strength. The operational principle of the farinograph is simple: the torque transmitted by the dough during mixing by two sigmoidal shaped paddles which rotate at constant speed is measured. As the interaction among the various flour components increases, the torque increases. According to the structural organization of the dough depicted in the previous paragraph, a dynamic situation ensues in the dough in which new interactions begin, while others cease. During the dough development stage, at any specific time increment, the net balance between these two mechanisms favors the increased number of link formation, which is reflected in the progressively increasing torque. If mixing continues past the maximum torque in the farinogram, elongation of the protein network between two subsequent entanglement zones occurs, resulting in a discontinuous network and decreased dough strength. However, proteins are still able to disentangle rapidly and the decrease in strength is limited. Additional mixing results in the protein network breaking down. During mixing, portions of the dough network are stretched (building internal stresses) and the proteins retract (relaxing part of these internal stresses). This mechanism is prevented, in part, by high viscosity, i.e., local friction, among the different parts of the dough. Mixing at low speeds after dough development reverses the mixing process, because the lower speed hinders protein extension while subsequently permitting the extended chains to retract to their preferred dimension.

Extensibility and resistance to extension of a dough is usually estimated by means of the extensograph. With this instrument, a piece of dough molded into a cylinder is stretched by a hook. The ultimate strength at which the dough collapses and the corresponding elongation is measured. Again, an explanation of the dough response during this test may be given in terms of the entangled protein network model. When a piece of dough is subjected to elongation, it will tear when the network between the two entangled regions reaches its full extension. The more entangled the network, the higher its resistance to deformation. Consequently, one would expect that flour containing highly branched proteins with very high molecular weight to show an increased maximum resistance, R_{max}. MacRitchie and Lafiandra[8] reported that R_{max}, estimated by means of the extensograph correlated poorly with the percentage of total polymeric proteins but was highly correlated with the percentage of unextractable polymeric proteins. These results demonstrate that entanglements, which contribute actively to the dough strength, are formed in the gluten network only at or above a critical molecular weight. Conversely, the extensibility of a dough correlates well to the total protein content of the flour and even better to the polymeric protein content. In order to elongate the dough, two different events must occur: the portion between the entanglement must extend by overcoming the friction between the proteins and the surrounding material, and the protein chains must slip through the entanglements. The rate at which these two events evolve governs the material response. If the rate of slippage is much faster than the rate of extension, the dough will elongate easily and fail, but if the rate of slippage is much slower than that of extension, the stress builds up suddenly and failure occurs rapidly. For proper deformation to occur, the dough structure should allow both mechanisms to occur at the same rate.

The capability of a dough to retain air bubbles is commonly estimated by using the alveograph. This simple test involves measuring the pressure change with time as air is blown at a constant rate below a disk of dough clamped to the instrument. The measure of pressure increment over time continues until the bubble collapse. The height of the peak, P, the length of the curve, L, and the area below the curve, W, are the parameters used to predict flour performance during leavening and oven spring. The parameter P is related to the dough's ability to retain gas bubbles, L predicts the handling characteristic of the dough, and W is the baking strength of the flour. Strong flours, which perform better during baking, have higher P and W values and an L value which is not too short. From the rheological point of view, this test is an approximate measure of the strain hardening of the dough during biaxial extension, which is considered the most important requirement for gas retention.[14] A small amount of air is occluded in the dough during the mixing operation, forming small spherical gas cells whose size increases during the fermentation stage as part of the carbon dioxide produced by the yeast migrates into them. In the early stage of baking, the cell's volume further increases due to the decrease in the solubility of the gas dissolved in the dough's aqueous phase and the subsequent changes in the physical state of volatiles. The dough becomes a foam with polyhedral gas cells. The dough film between the cells is tangentially extended in two directions and compressed radially. To prevent the rupture of the bubbles and subsequent loaf collapse, it is essential that the resistance against extension at the thinning section of the cell wall is greater than that at the thicker portion of the film, allowing the point of weakness in the bubble wall to be repaired. From this point of view, the existence of entangled regions in the gluten network is of fundamental importance. As the biaxial shear rate increases, slippage of the protein chains through the entanglements is hindered. The entanglements then act as a permanent constraint, allowing resistance to elongation to increase disproportionately with increased strain.

During baking, the dough undergoes severe irreversible changes. These changes are highly dependent on the heat transfer rate from the oven to the loaf and the corresponding rheological behavior of the dough. Starch modifications and protein starch interactions during this stage also help transform the dough from a viscoelastic liquid to a semi-rigid solid. As the dough temperature rises, starch granules begin to swell (~40°C). Before swelling becomes significant, dough fluidity increases due to enzymatic degradation of starch by amylases and mobility enhancement of the gluten network. During swelling, the starch granules compete with proteins for water absorption from the liquid phase. However, due to the limited supply of water, a large portion of the granules remains intact. Once the crumb temperature reaches 60 to 70°C, the gluten proteins undergo thermal denaturation, significantly reducing their water binding capacity, and more starch granules gelatinize. Additionally, intact granules become flexible and can elongate, promoting thinning of the gluten film as the gas in the cell expands. However, by this time, starch gelatinization has advanced sufficiently, preventing collapse of the dough structure.

Considering this complex mechanism, it is not surprising that tests performed at room temperature are unable to fully predict the baking performance of the dough. Alternatively, estimations of baking performance may be obtained by measuring the loaf volume increment during baking[15] or by fundamental-type tests which characterize the evolution of the dough rheological behavior with varying temperature.[4,16-19]

FUNDAMENTAL TESTS

Although the imitative tests have been used for a long time as standards for evaluating flour performance, their interpretation is hampered since the parameters estimated are often a reflection of the geometry and dynamics of the testing apparatus and the result of stretching and flow properties of the dough. Consequently, these tests may help in differentiating the behavior of doughs with different compositions, but they are not useful for elucidating the functional role played by the constituents of the various flours and the nature of their interactions. Similarities between the rheological behavior of flour dough and polymeric materials of biological and non-biological origins have been recognized. Consequently, further understanding of dough rheology may benefit from theories and investigative techniques successfully used in synthetic polymer science to interpret the behavior of bio–polymers.

Viscoelasticity of a dough can be investigated using dynamic methods. These methods provide simultaneous assessment of the elastic and viscous components and are used frequently to study flour doughs. In these measurements, a sample is placed between two parallel plates or between a plate and a cone of very small angle, and is subjected to sinusoidal displacement. The force generated by the motion of one plate is transmitted through the sample to the other plate and recorded. If the material between the plates is a perfectly elastic material, Hook's law governs the stress-strain relationship and consequently stress and strain wave functions will have a 0° lag (in phase). Conversely, if the material is a viscous liquid, Newton's law applies, and the lag between stress and strain waves will be 90° (out of phase). A viscoelastic material will have a lag angle, δ, between 0 and 90°. Complex modulus, G^*, is the ratio between stress wave and strain wave amplitude, and it can be separated into two components: 1. an elastic (in phase), or stored component with the strain (G'), and 2. a viscous (out of phase) G'', or dissipated component. tan δ is calculated from G''/G' and is the measure of the relative contribution of each component. Dynamic tests are commonly performed by varying the angular frequency of the strain wave in a wide range (2 to 3 decades) observing the material response at different time scale. A curve of G' and G'' versus frequency, ω, allows one to obtain information on the relaxation time spectrum, which is an inherent characteristic of the viscoelastic materials.[20] While curves of tan δ versus ω can help in detecting occurrence of conformational changes as they present a maximum in correspondence of these events. Figure 1 shows the results obtained when wheat flour doughs are submitted to dynamic tests at room temperature. Both the elastic and viscous contributions to the complex modulus increase with increasing the frequency, i.e., with reducing the time scale of observation. Although the relative weight of each parameter contribution varies (tan δ), the elastic character of the dough always dominates (tan $\delta < 1$).

To relate this information to dough structure, the sample deformation should not involve structural damage. Data illustrated in Figure 2 represents the dynamic test response resulting from increasing the strain amplitude. Below a certain critical value (~0.2% strain) the rheological response is independent of the strain. Above this critical strain level, the resistance against deformation is progressively reduced and the viscous character of the dough becomes more relevant. Different critical

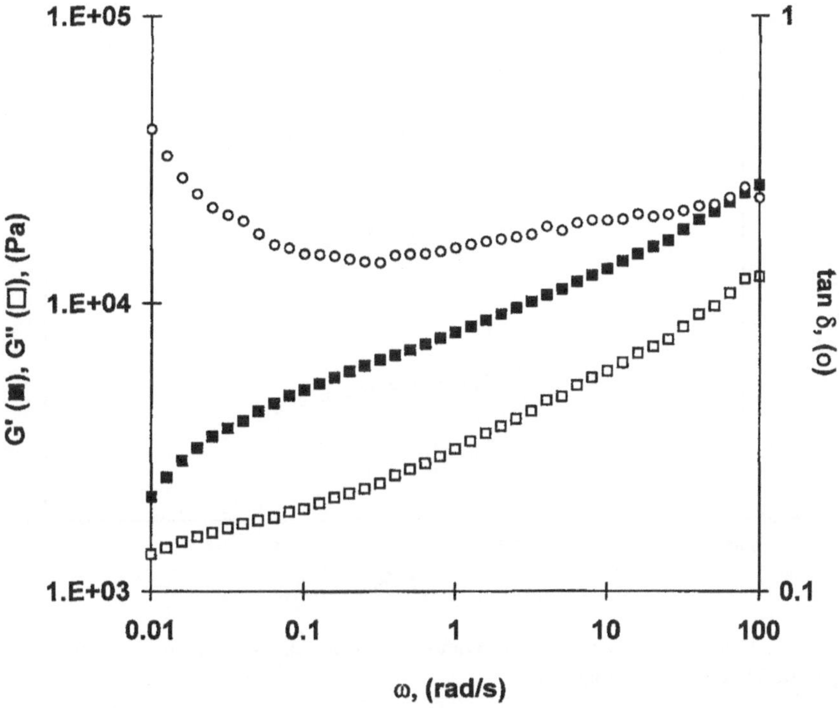

FIGURE 1 Storage modulus (G′), loss modulus (G″), and tan δ of wheat flour dough versus frequency, ω (rad/sec) at room temperature.

strain levels for dough have been reported,[21] but this should be not surprising, as these depend on dough composition and water content.

With new instrumentation that is available, it is now possible to more accurately control the sample temperature during an experiment. Dynamic mechanical testing can be used to monitor the baking behavior of different doughs and the mode of action of active compounds or additives.[22] Figure 3 shows how the dynamic properties evolve during this type of test. During the earliest stage of the heating process, both G′ and G″ decrease as the dough fluidity increases. At higher temperatures, G′ and G″ increase as much as two orders of magnitude, partly due to starch gelatinization and protein denaturation. The transition of dough behavior from a viscoelastic liquid to a semi-rigid solid can be observed from changes in the tan δ, which approaches values near zero corresponding to elastic materials.

Performing dynamic experiments on doughs at very low frequency is limited by the length of the experiment. To obtain reproducible results, each experiment should last long enough to observe the material response over three or four deformation cycles. Consequently, at frequencies below 10^{-2}, an experiment may last for hours or even days. During this time the sample undergoes to severe modification due to drying and fermentation, which makes the result questionable. A stress relaxation test or creep test may be more appropriate to gain structural information over a longer time period. In a stress relaxation experiment, a constant strain (γ_0) is applied

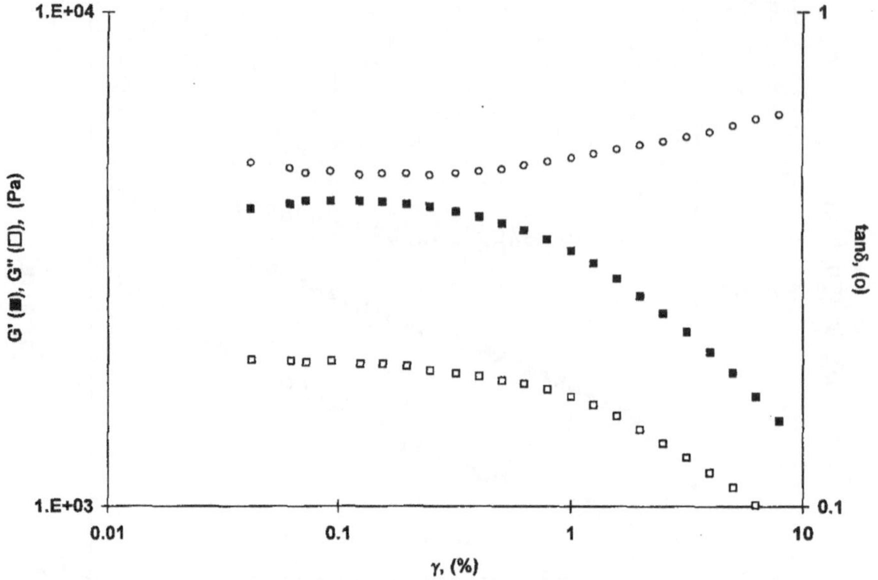

FIGURE 2 Storage modulus (G′), loss modulus (G″), and tan δ of wheat flour dough at room temperature versus strain amplitude.

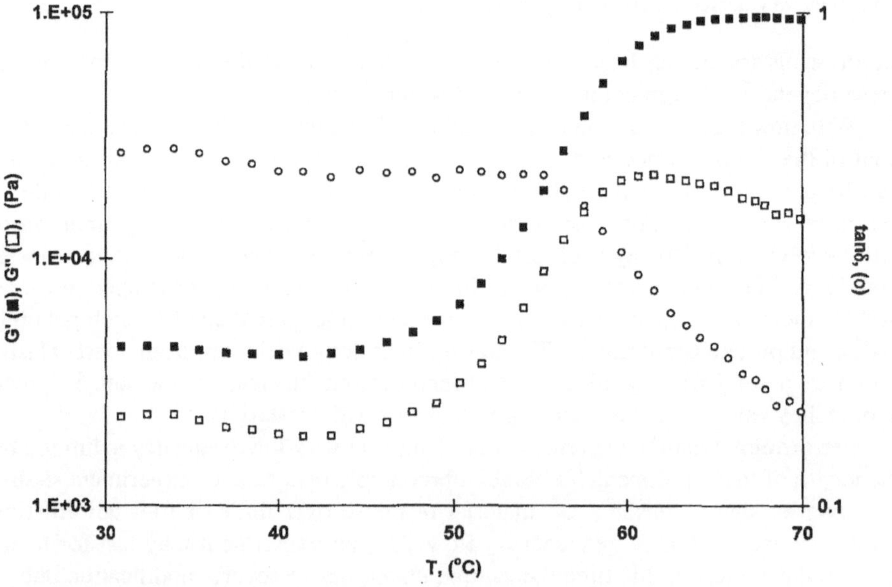

FIGURE 3 Storage modulus (G′), loss modulus (G″), and tan δ of wheat flour dough versus temperature.

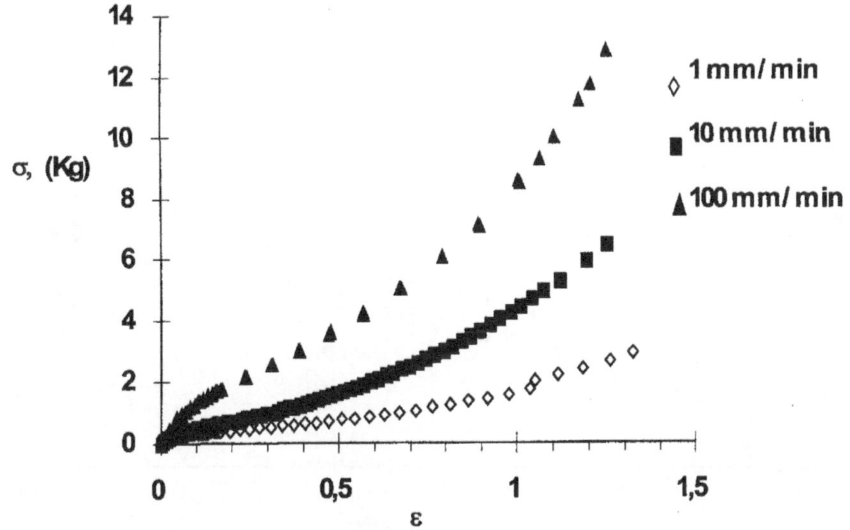

FIGURE 4 Compressive stress, σ, as a function of Henky's strain, ε, in lubricated squeezing test at different cross-head speeds, mm/min: (◇) = 1; (□) = 10; (▲) = 100.

to the sample, and the resulting stress ($\sigma(t)$) is measured as a function of time. The time-dependent modulus, or E(t), is described as the $\sigma(t)/\gamma_0$ ratio. In creep experiments, a sample is subjected to a constant stress (σ_0), and the resulting strain ($\gamma(t)$) is measured as a function of time. Time-dependent compliance, or J(t), is calculated as the $\gamma(t)/\sigma_0$ ratio.

Other fundamental-type tests have recently proven valuable in characterizing rheological behavior of doughs.[23] Among these, the lubricated squeezing test can be used to observe dough response in elongational extension. This test is performed by applying uniaxial compression to a sample placed between lubricated parallel plates of a dynamometer. The sample biaxial extensional flow is evaluated to estimate the elongational viscosity of the dough.[14,24] An example of the stress-strain curve generated in lubricated squeezing experiments at different cross-head speeds is shown in Figure 4.

THE VISCOELASTIC BEHAVIOR OF DOUGH AS OBSERVED WITH FUNDAMENTAL TESTS

THE INFLUENCE OF THE PROTEIN FRACTION

Fundamental tests have been used by many authors to study the rheological behavior of dough made of different flours; however, the interpretation of these results must be looked at carefully. For example, various authors[6,25] have attempted to correlate the physical properties estimated from such tests with parameters derived from empirical results. Operating in the linear viscoelasticity range, it was not possible to distinguish between two types of doughs. However, by performing the test at

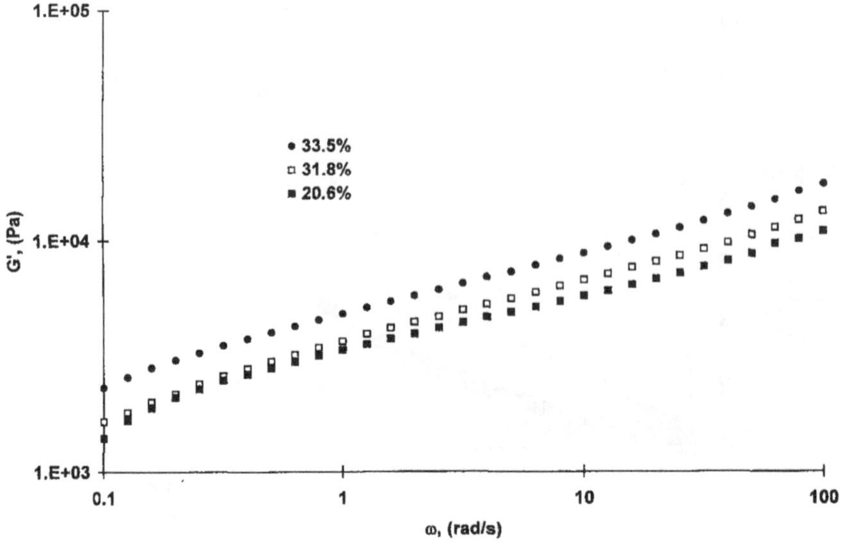

FIGURE 5 Storage modulus (G′) versus frequency ω (rad/sec). of doughs having different gluten contents (w/w): (●) = 33.5; (□) = 31.8; (■) = 20.6.

high-strain amplitude, i.e., outside the linear viscoelasticity range, a correlation between viscoelastic properties of the doughs and their strength could be found by the farinograph. These results are, apparently, in contrast with the previous recommendation for experimental conditions which are not destructive to the dough structure. In this case, fundamental tests were used improperly, since these are designed to obtain the structural organization of the material rather than its performance in different situations. The fundamental approach is superior to the empirical one for the purpose of obtaining structural organization. When the sample is subjected to small deformations, the response is related to the kinetics of conformational changes that evolved during the shearing period. Thus inter- and intramacromolecular interactions can be derived. In a large-deformation test, the dough structural network is progressively damaged, and the response is completely different from that previously described. In this case, the response is more sensitive to the damage caused during deformation than to the structural organization of the dough. To better illustrate this aspect, Figure 5 shows the dynamic storage modulus of dough prepared from flour of different origins (i.e., gluten content) but with the same water content. A typical relationship between the storage modulus curves and the flour proteins content can be observed. The G′ curves are parallel in the high frequency region, but they begin to diverge in the low frequency region. The stress-relaxation curves of the same dough shown in Figure 6 further support this hypothesis. This data suggests that distinguishing one flour from another is better accomplished by observing the conformational changes, which require a longer time to evolve.

Interesting results have been found when the molecular origins of dough rheology were studied. Petrofsky and Hoseney[26] examined the behavior of starch and gluten mixtures in dynamic tests, observing that hard wheat gluten doughs showed low G′

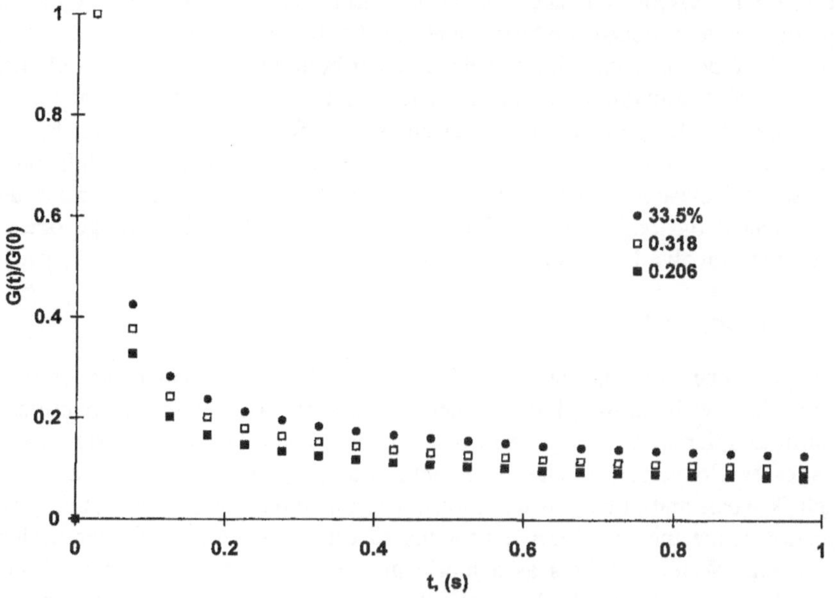

FIGURE 6 Stress-relaxation curves of doughs having different gluten content (w/w).

and G″ values, while soft wheat doughs showed higher G′ and G″ values. They hypothesized that the two different flours interacted in a different way with starch. Their conclusion may explain why soft wheat doughs are less extensible than hard wheat doughs. By adding the appropriate chemical species to a dough, it is possible to inhibit or enhance molecular interactions among the dough constituents and thus obtain direct information on their functional roles. For example, the addition of ascorbic acid has been shown to cause both the storage and loss moduli to increase in agreement with this compound's well-known strengthening effect. However, tan δ did not vary significantly. Conversely, an addition of glutathione resulted in a decrease of G′ and G″ as well as a change in tan δ.[27] Similar results were obtained by analyzing the viscoelastic behavior of the gluten alone. Addition of deuterium oxide did not cause conformational changes in the gluten, while gluten reconstituted with urea showed a reduced frequency dependence of the storage modulus and a large decrease in the magnitude of the loss modulus, especially at the low frequency region.[28] Dong and Hoseney[29] demonstrated that the major factor causing the change in rheological behavior of a dough was the sulfhydryl-disulfide interchange reaction. Addition of potassium bromate to a dough resulted in an increase in G′ and a decrease in loss tangent (more elastic) while the addition of glutathione reduced G′ and increased the loss tangent (less elastic). According to a network entanglement model used to describe the structural organization of dough,[30] gluten structure is mainly responsible for maintaining the elasticity of a dough at least partly due to disulphide crosslinks existing in the network formed by the protein molecules. The storage modulus, which is a measure of the elastic character of the dough, is expected to increase with increasing number of disulphide bonds. The loss modulus, which is a measure of the viscous character, is expected to increase with higher low-molecular sulfhydryl content. Potassium bromate removes sulfhydryl groups by oxidizing them to disulphide bonds, enhancing the elastic contribution to the dynamic modulus. On the other hand, glutathione acting as a reductant, increases the rate of the thiol-disulphide interchange reaction. Fewer cross-links form, and as a result the size of large proteins decreases and the relative amount of low molecular weight proteins increases. All these results support the idea that rheological behavior of a dough is more affected by long range conformational changes than short range ones, i.e., those which involve large portions of the protein network.

The Influence of Water

Water is of primary importance in determining the rheological behavior of flour doughs. It is well known that if water is added to flour in small quantities, the resulting dough may have insufficient cohesiveness and is not able to sustain mechanical stresses. Conversely, if a large amount of water is added to the flour, the resulting dough is weak and sticky. Water molecules are important in the gluten network formation since they enable the proteins to establish intra- and inter-molecular interactions. Water also acts as a lubricant, occupying the space in between the various flour components. In dough rheology, it is still not clear to what extent water behaves as an inert filler or to what degree it plays a more active role. Several authors have examined the rheological behavior of doughs with varying moisture content by means of dynamic tests.[27,31-35] Within a range of frequencies, both dynamic moduli (G' and G'') varied proportionally with water content but were independent of frequency. Moreover, the derivatives of log G' and log G'' were functions of frequency but independent of water content. It was concluded that the effect of frequency and water might be separated, and that water did not affect macromolecular mobility. However, this simplistic conclusion remains to be further elucidated.

In polymer systems, macromolecular mobility is governed by the transition from a glassy to a rubbery state which occurs at a temperature range, T_g, that is a characteristic property of the material. At temperatures below the T_g, the molecular mobility at short and long range is hampered, while at temperatures above T_g, large-scale rotational and translational motion of macromolecules occurs, allowing molecular reorganization phenomena to become detectable in the experimental time scale. Levine and Slade[36] showed experimental evidence that water acts as a ubiquious plasticizer in many foods by lowering their T_gs. Data concerning the T_g range of flour doughs with varying moisture contents are not available in the scientific literature. However, it has been reported that when moisture content in gluten as well as in glutenins was raised above 14 to 16%, glass transition decreased to below room temperature.[37,38] Therefore, it is possible that at ambient temperatures, the gluten network of a dough is in the rubbery state, and thus, long range and short range molecular reorganization processes should be detectable on the experimental time scale. The fact that an influence of water on the polymer molecular mobility was not observed may be due to use of inappropriate frequencies in the dynamic tests performed. By exploring a wider range of frequencies the influence of water on the macromolecular mobility of the dough may be observed.[39] Moreover, the

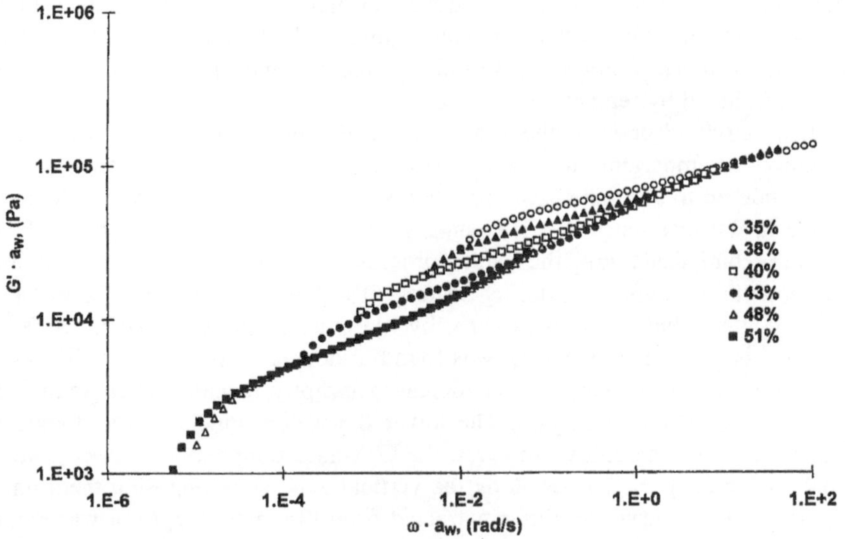

FIGURE 7 Reduced storage modulus, $G' a_D$, master curve versus reduced frequency, ωa_C. Moisture content (w/w) as shown in the figure.

behavior of a concentrated polymer system such as dough can be characterized by using a double reduction procedure: shifting the rheological curves along the vertical axis to account for the dilution effect of water, and shifting the curves along the horizontal axis to account for the influence of water on T_g. This procedure generates a master curve that can be considered characteristic of the viscoelastic behavior of the dough.

Figure 7 shows more recent unpublished data obtained by further lowering the experimental frequency range. The double reduction procedure does not generate a master curve, but instead a family of curves which deviate from the master curves in the very low frequency region. This new experimental procedure has led to a hypothesis that water not only affects macromolecular mobility, but also interferes with the mechanisms with which molecular conformation changes over time. Stress relaxation experiments performed on the same flour doughs also supported this conclusion. Stress decay curves (not shown) were fitted by using a three-element generalized Maxwell model,[20] providing an estimate of the main relaxation times. By increasing the moisture content from 45.5 to 56.5%, the relaxation time was reduced from 120 s to 78 s. If water molecules act as an inert filler, differences should occur in the relaxation modulus, but the decay kinetics should remain the same.

During the baking process, rheological behavior of a dough changes significantly as a consequence of protein denaturation and starch gelatinization. Water availability is important in these processes and the resulting rheological behavior. When gluten is dynamically analyzed at a constant rate and increasing sample temperatures, its G' decreases slightly, up to 60 to 70°C. Above 70°C, gluten starts to polymerize and G' increases. If the water content of the gluten-water mixture is varied, both G' and G'' change, while the tan δ remains the same. A vertical shift of the curve results in a single curve for both G′ and G″.[18] These experimental observations suggested that excess water at a level beyond the amount required to hydrate the gluten (c.a. 33%)[40] exerted only a lubrication effect and did not interfere with gluten structural transformation induced by temperature change.

When starch is present, the shape of the dynamic property curves changes completely. At temperature well below 70°C, G′ increased by more than two orders of magnitude up to a broad maximum, then slightly decreased, corresponding with a change in the loss tangent.[35] This suggests that interactions between starch and protein are established during the heating process. Evolution of the elastic component of the dynamic modulus, G′, during heating of a dough containing 45.8 to 54.4% moisture contents has been investigated by Marchisano and coworkers.[41] A family of curves was generated, where G′ was found to decrease up to 40 to 50°C. As the temperature was further increased, G′ increased abruptly, reaching a maximum value near 70°C, then decreased again. The lower the water content of the dough, the greater its dynamic modulus. However, the G′ versus temperature curves were not parallel. By shifting the curves along the vertical axis (superimposing them on the first portion of the curve for gluten extracted from the same dough), it was evident that the difference between the curves for dough and gluten could be attributed to starch-water interactions, and these interactions increased as water content in the dough increased. In addition, the interaction below 39% moisture was estimated to be negligible from the rheological point of view. Since the amount of water in the dough determines the occurrence and extent of starch gelatinization, the rheology of the dough is affected during baking. If the moisture content is lower than 39%,[42] starch gelatinization does not take place and it acts like an inert filler with respect to rheology of dough. Above this moisture content, water promotes starch gelatinization, which in turn is largely responsible for the enhancement of mechanical strength of the dough during baking.

The Influence of Other Flour Constituents

Dough is a composite of materials formed by two continuous phases: the swollen protein network, and the liquid phase which surrounds it, in which starch granules are dispersed. The rheological behavior of doughs is mainly due to interactions of proteins in gluten. Cavella and coworkers[17] reported that dough elasticity decreased with starch content, resulting in an increase in the viscous component. This was further explained as a result of a decrease in volume concentration of protein with increasing starch content, and consequently, a looser and less elastic network formed after hydration. In contrast, starch granules, which do not take part in the network formation, may enhance viscous dissipation during motion. Therefore, one may hypothesize that starch, in the dough development stage at room temperature, plays a less important role than gluten, in the dough structure-forming process where it acts primarily as a filling material. However, gluten-starch interaction has been postulated to be of importance in determining the viscoelastic behavior of the dough.[26,43] This hypothesis was based on some experimental evidence where starches isolated from different wheat cultivars were mixed into a dough with a constant gluten level, resulting in significant rheological differences. In particular, soft wheat and non-wheat starch doughs had higher moduli than hard wheat starch doughs, as a consequence of greater interaction between starch and gluten. Low G′ and G″ values in the hard wheat doughs indicated less starch-gluten interaction. This discrepancy merits further investigation, as it may be relevant in predicting the bakery quality of a flour.

Lipids also exert some influence on the rheological properties of the doughs. Added lipids lowered G′ and G″ in the rubbery region at 25°C. In stress relaxation tests, the longer relaxation time was found to increase with higher lipid content, simultaneously delaying the onset of viscous flow and attenuating the elastic properties of the gluten fraction.[44] By comparing the rheological behavior of whole flour dough and lipid-less flour dough,[45] lipids were found to play a role in the gluten network formation. A possible explanation is that lipids decreased the availability of water molecules during the dough development process, and consequently lipid-less gluten-starch-water mixtures reacted faster, resulting in a more consistent dough than the corresponding mixture prepared with lipid-less gluten.

CONCLUSION

Full understanding of the rheological behavior of flour dough is of great importance from the practical point of view. Dough rheology directly affects the baking performance of flours, and rheological analyses have been made in order to optimize dough formulation. Although dough rheology has long been investigated, there remains a significant lack of understanding. This lack of progress is due to the complexity of this biological system and to inadequate instrumental approaches in molecular structure analysis in the past. The aim of this review was to illustrate some fundamentals of a novel approach based on the similarity between flour doughs and polymer materials, using fundamental-type rheological tests. This subject is in its infancy and only lately has sparked the interest of technologists and scientists who work in dough rheology. Like other novel approaches, it provides only a partial understanding and sometimes contradictory results. With increased experience and further in–depth investigations, subsequent data obtained should provide better understanding of dough rheology on a molecular basis.

REFERENCES

1. Scott Blair, G. W., *Foodstuffs, Their Plasticity, Fluidity, and Consistency,* Interscience, New York, 1953.
2. Schofield, R. K. and Scott Blair, G. W., The relationship between viscosity, elasticity and plastic strength of soft materials as illustrated by some mechanical properties of flour doughs, *Proc. R. Soc. London,* 138, 707, 1932.
3. Chung, O. K., Lipid-protein interactions in wheat flour, dough, gluten, and protein fractions, *Cereal Foods World,* 31, 242, 1986.
4. Attenborrow, G., et al., Rheological properties of wheat gluten, *J. Cereal Sci.*, 12, 1, 1990.
5. Castel-Perez, E. M. and Steffe, J. F., Viscoelastic properties of dough, in *Viscoelastic Properties of Foods*, Rao, M.A. and Steffe, J.K., Eds., Elsevier, London, 1992, chap. 3.

6. Lindborg, K. M., Tragardh, C., Eliasson, A.–C., and Dejmek, P., Time-resolved shear viscosity of wheat flour doughs — effect of mixing, shear rate, and resting on the viscosity of doughs of different flours, *Cereal Chem.*, 74, 49, 1997.
7. Bloksma, A. H., Dough structure, dough rheology, and baking quality, *Cereal Foods World*, 35, 228, 1990.
8. MacRitchie, F. and Lafiandra, D., Structure-function relationship of wheat proteins, in *Food Proteins and Their Applications*, Damadaran, A. and Paraf, S., Eds., Marcel Dekker, New York, 1997, chap. 10.
9. Slade, L. and Levine, H., Beyond water activity: recent advances based on an alternative approach to the assessment of the food quality and safety, *Crit. Rev. Food Sci. Nutr.*, 30, 115, 1991.
10. Ewart, J. A. D., A modified hypothesis for the structure and rheology of glutenin, *J. Sci. Food Agric.*, 19, 617, 1972.
11. MacRitchie, F., Physicochemical processes in mixing, in *Chemistry and Physics of Baking*, Blanshard, J.M.V., Frazier, P.J., and Galliard, T., Eds., RACI, London, 1986, 132.
12. Larsson, H. and Eliasson, A.-C., Phase separation of wheat flour dough studied by ultracentrifugation and stress relaxation. I: Influence of water content, *Cereal Chem.*, 73, 18, 1996.
13. Walker, C. E. and Hazelton, J. L., Dough rheological tests, *Cereal Foods World*, 41, 23,1996.
14. Vliet, T., Janssen, A. M., Bloksma, A. H., and Walstra, P., Strain hardening of dough as a requirement for gas retention, *J. Texture Stud.*, 23, 439, 1992.
15. Ross, A. S. and MacRitchie, F., Interactions of wheat proteins, carbohydrates, and lipids, in *Ingredient Interactions, Effects on Food Quality*, Gaonkar, A. G., Ed., Marcel Dekker, New York, 1995, 321.
16. Masi, P., Study of the influence of temperature on the rheological behaviour of gluten by means of dynamical mechanical analysis, in *Food Properties and Computer Aided Engineering of Food Processing Systems*, Singh, P. and Medina, A., Eds., NATO ASI Series, Kluwer Academic, London, 1989, 357.
17. Masi, P., Cavella, S., and Piazza, L., Mechanical dynamical analysis of gluten undergoing thermal denaturation, in *Engineering and Food,* Vol. 1, *Physical Properties and Process Control*, Spiess, W. and Shubert, H., Eds., Elsevier Science, Barking, England, 1990, 122.
18. Cavella, S., Piazza, L., and Masi, P., Thermomechanical analysis of gluten-starch-water mixtures. Influence of testing conditions and composition, *Ital. J. Food Sci.*, 2, 235, 1990.
19. Kokini, J. L., Cocero, A. M., Madeka, H., and de Graaf, E., The development of state diagram for cereal proteins, *Trends Food Sci. Technol.*, 5, 281, 1994.
20. Ferry, J. D., *Viscoelastic Properties of Polymers,* 3rd ed., John Wiley & Sons, New York, 1980.
21. Weipert, D., The benefits of basic rheometry in studying dough rheology, *Cereal Chem.*, 67, 311, 1990.
22. Weipert, D., A recording baking test, *Bäker Konditor*, 37, 167, 1988.
23. Morgenstern, M. P., Newberry, M. P., and Holts, S. E., Extensional properties of dough sheets, *Cereal Chem.*, 73, 478, 1996.
24. Bagley, E. B., Christianson, D. D., and Martindale, J. A., Uniaxial compression of hard wheat flour dough: data analysis using the upper convected Maxwell model, *J. Texture Stud.,* 19, 289, 1988.
25. Safaril-Ardil, M. and Phan-Thien, N., Stress relaxation and oscillatory tests to distinguish between doughs prepared from wheat flours of different varietal origin, *Cereal Chem.*, 75, 80, 1998.

26. Petrofsky, K. E. and Hoseney, R.C., Rheological properties of dough made with starch and gluten from several cereal sources, *Cereal Chem.*, 72, 53, 1995.
27. Berland, S. and Launay, B., Rheological properties of wheat flour doughs in steady and dynamic shear: effect of water contents and some additives, *Cereal Chem.*, 72, 48, 1995.
28. Inda, A. E. and Rha, C., Dynamic viscoelastic behaviour of wheat gluten: the effect of hydrogen bonding modification by urea and deuterium oxide, *J. Texture Stud.*, 22, 393, 1991.
29. Dong, W. and Hoseney, R. C., Effects of certain breadmaking oxidants and reducing agents on dough rheological properties, *Cereal Chem.*, 72, 58, 1995.
30. Bloksma, A. H. and Bushuk, W., Rheology and chemistry of dough, in *Wheat Chemistry and Technology*, Vol. 2, Pomeranz, Y., Ed., American Association of Cereal Chemistry, St. Paul, MN, 1988.
31. Hibberd, G. E. and Wallace, W. J., Dynamic viscoelastic behavior of wheat flour doughs. I: Linear aspects, *Rheol. Acta*, 5, 193, 1966.
32. Hibberd, G. E., Dynamic viscoelastic behavior of wheat flour doughs. II: Effect of water content in the linear region, *Rheol. Acta*, 9, 497, 1970.
33. Hibberd G. E. and Parker, N. S., Dynamic viscoelastic behavior of wheat flour doughs. IV: Non linear behaviour, *Rheol. Acta*, 14, 151, 1975.
34. Navickis, L. L., Anderson, R. A., Bagley, E. B., and Jasberg, B. K., Viscoelastic properties of wheat flour doughs: variation of the dynamic moduli with water and protein content, *J. Texture Stud.*, 13, 249, 1982.
35. Dreese, P. C., Faubion, J. N., and Hoseney, R. C., Dynamic rheological properties of flour, gluten, and gluten-starch doughs. II: Effect of various processing and ingredients changes, *Cereal Chem.*, 65, 354, 1988.
36. Levine, H. and Slade L., Glass transition in foods, in *Physical Chemistry of Foods*, Schwartzberg, H. G. and Hartel, R. W., Eds., Marcel Dekker, New York, 1992, 83.
37. Hoseney, R. C., Physical chemistry of bread dough, in *Physical Chemistry of Foods*, Schwartzberg, H. G. and Hartel, R. W., Eds. Marcel Dekker, New York, 1992, 443.
38. Madeka, H., and Kokini, J. L., Change in rheological properties of gliadin as a function of temperature and moisture: development of a state diagram, *J. Food Eng.*, 22, 241, 1993.
39. Masi, P., Cavella, S., and Sepe, M., Characterization of dynamic viscoelastic behavior of wheat flour doughs at different moisture content, *Cereal Chem.*, 75, 428, 1998.
40. Lasztity, R., *The Chemistry of Cereal Proteins*, CRC Press, Inc., Boca Raton, FL, 1984.
41. Marchisano, C., Gennaro, L., Sepe, M., and Masi, P., Characterizing *in situ* starch gelatinization. Thermal and dynamic mechanical analysis of durum wheat dough, *J. Thermal Anal.*, 46, 181, 1996.
42. Eliasson, A.-C. and Gudmundsson, M., Starch: physicochemical and functional aspects, in *Carbohydrates in Food*, Eliasson, A.-C., Ed., Marcel Dekker, New York, 1996, 431.
43. He, H. and Hoseney, R. C., Factors controlling gas retention in non heated doughs, *Cereal Chem.*, 69, 1, 1992.
44. Fu, J., Mulvaney, S. J., and Cohen, C., Effect of added fat on the rheological properties of wheat flour doughs, *Cereal Chem.*, 74, 304, 1997.
45. Cavella, S., Masi, P., Chianese, L., Petrosino, R. M., and Sacchi, R., The functional role of durum wheat lipids during the pasta manufacturing process, in *Proceedings ICC Symposium: Cereal Based Foods: New Developments*, Prague, Czechoslovakia, 1991, 456.

5 Instrumental Techniques Used in Bread Staling Analysis

Yael Vodovotz, Mooyeol Baik, Elena Vittadini, and Pavinee Chinachoti

CONTENTS

INTRODUCTION

Bread staling is characterized by many physical and chemical phenomena such as changes in texture, water migration, starch crystallization, and component interactions. Analyses of bread staling have evolved over time, as instrumental technology and knowledge of the subject have progressed. Various analytical techniques are available for monitoring changes at macroscopic, microscopic, and molecular levels. Careful selection of techniques offers a strategy to investigate the bread staling mechanism. It is imperative to consider the various events taking place concurrently

0-8493-8790-6/01/$0.00+$.50

within the system. Each event occurs in a distinct time frame and scale (e.g., macroscopic to molecular). Using appropriate instrumental techniques, useful data can be obtained leading to a better understanding of the nature of bread staling. Examples of such studies applied to bread can be found elsewhere.[1-5]

MACROSCOPIC PROPERTIES

Sensory tests are most common in the practical world for evaluating the staleness of bread. However, sensory evaluation suffers from subjective results that are difficult to relate to physiochemical phenomena probed by instrumental analysis. Of all the physical and chemical properties of bread, mechanical properties have received the greatest attention, since they relate to textural perception. Texture properties are of paramount importance in elucidating macroscopic changes in quantitative terms. Thereby, they enable the quantitative evaluation of the cellular structure of bread. Imitative tests for dough performance were covered in detail in Chapter 4. Therefore, this chapter will focus on the fundamental tests used to understand the mechanism of bread staling.

TEXTURE

Bread is a spongy material which tends to harden and crumble upon staling. These changes can be characterized using mechanical theories specifically developed for solid foams and cellular solids.[6] Mechanical properties such as firmness at a given compressive strain, compressive stress-strain relationships, recoverable work, and tensile strength are evaluated.

Firmness can be described as the force required to compress breadcrumb to a given deformation. It is the most popular method of evaluating staleness of bread. Firmness has been shown to strongly correlate with sensory hardness,[7] and hence has served as the most common analytical tool to assess bread staling.[7-10]

Various instruments are available to measure firmness, including the Instron Universal Testing Machine (IUTM), Baker Compressimeter (BC), Texture Analyzer (TA), and Precision Penetrometer (PP). Since so many devices and methods are available, a comparative evaluation was undertaken[11] which failed to identify a universally acceptable method for measuring the physical properties of crumb.[12] Performance comparison among these instruments has been reported.[13]

Textural changes upon staling have been evaluated using the Avrami analysis,[14-19] and examples are described in Chapter 9. Typical Avrami exponential constants of breads and starches have been calculated to be close to one.[19] These results suggest that the mechanism of starch retrogradation is instantaneous nucleation followed by rod-like growth of crystals,[19] and that bread staling involved recrystallization of starch in the crumb. However, other researchers reported that the Avrami exponential constant was less than one during the first 24 hours of storage, indicating an additional firming process during this time frame which is superimposed on the crystal growth in the starch fraction.[17,20]

Since bread has a spongy structure, its compressive behavior can be studied in light of mechanical theories specifically developed for solid foams and cellular solids. One of the popular methods used to investigate the mechanical properties of bread is the characterization of compressive stress-strain relationships. Ashby[21] suggested theoretical relationships between the structure of cellular solids and their mechanical properties. In ideal spongy material, which has elastic cell-wall and uniform cell size, the compressive stress-strain curve has three clearly identified regions: elastic or quasi-elastic deformation as a result of cell-wall bending, collapse as the cell-wall buckles, and yield or fracture and densification as a result of cell-wall crushing (Figure 1).[21,22] Although bread is not an ideal spongy material, the analysis provides a quantitative indication as to the role of the cellular structure in determining the mechanics of solid foams. In sponge cake, the magnitude of the initial modulus, i.e., the slope of the first part of the engineering stress-strain curve, was found to be proportional to its bulk density.[23] Several researchers explained the characteristics of three different regions of sigmoidal compressive stress-strain curve of spongy baked goods by using the empirical models.[6,24,25]

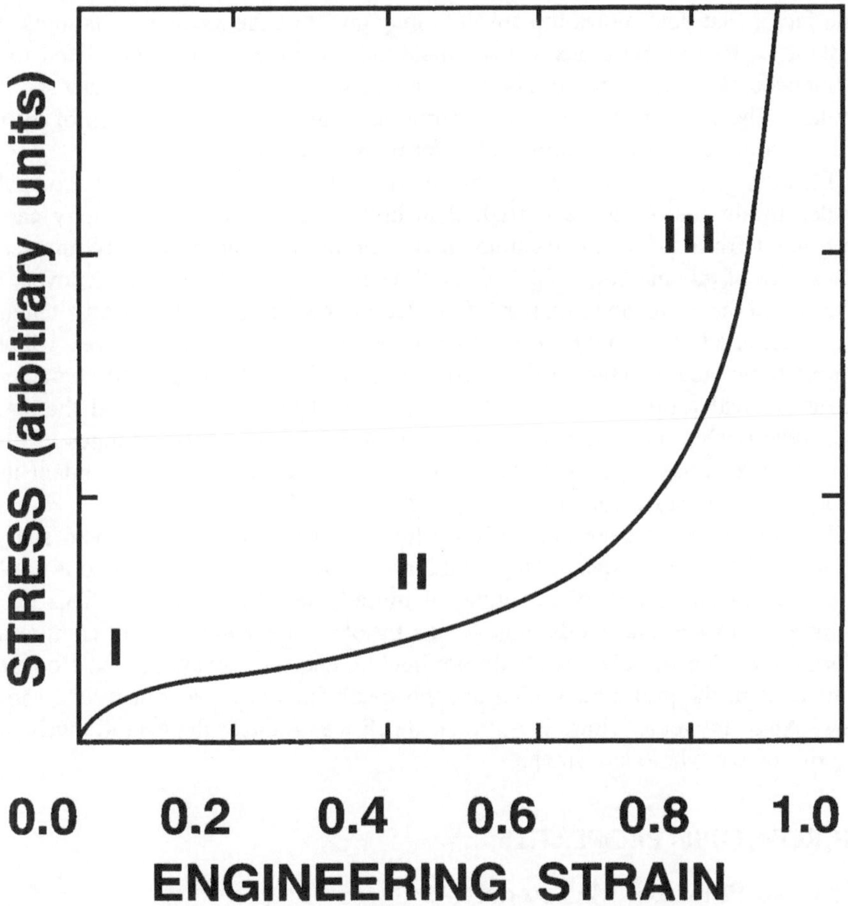

FIGURE 1 Typical sigmoid shape of the compressive stress-strain relationship of cellular solids. I. Small, primary elastic deformation. II. The 'shoulder' representing buckling and/or fracture of cell walls. III. The rapid stress increase when collapsed cell wall material is being compressed.[22]

Peleg and coworkers[6] described the compressive stress-strain curve using a single mathematical model with three empirical parameters which enabled the expression and quantification of changes in stress-strain relationships in terms of this model's constants.

$$\sigma = \frac{C_1 \varepsilon_E}{(1 + C_2 \varepsilon_E)(C_3 - \varepsilon_E)} \quad (1)$$

$$\varepsilon_E = \Delta H / H_0 \quad (2)$$

where σ is the stress, ε_E the engineering strain, ΔH the absolute deformation, and H_0 the initial specimen height, and C_1, C_2, and C_3 are constants. These three constants represent overall rigidity of the material, height of the flat region of the stress-strain curve, and strain level at which densification occurred.[6] They have the following relation to the shape of the compressive stress-strain curve. The constant C_1 is a scale factor that determines the absolute magnitude of the stress and its units. The constant C_2 is a shape index whose magnitude relative to unity is related to the prominence of the shoulder, but both C_2 and C_3 actually describe the exact shape of shoulder. The constant C_3 reveals the strain level at which densification of the cell wall material becomes the dominant deformation mechanism.[6]

Elasticity (the ability of a body to return to its original shape after removal of the deforming load) has been studied in bread. The degree of elasticity can be determined from the residual strain in one or more compression-decompression cycles. Use of this method is fairly difficult because some of the strain recovery can be slow and the corresponding part of the decompression curve flat.[22] An alternative is to determine the area under the compression and decompression curves, which is also related to recoverable work. Recoverable work has been used to characterize the mechanical properties of bread.[25-29] Kou and Chinachoti[25] applied the stress-strain relationship and recoverable work to the study of textural changes of bread upon storage. They suggested that recoverable work was highly dependent upon storage time, moisture content, and deformation level.

Few reports have been published on the use of tensile tests for bread analysis due to the difficulty in specimen preparation.[30-32] In tensile tests, a force is applied which causes elongation of a sample of initially uniform geometry. This tensile deformation of cellular solids is governed by other mechanisms than compressive deformation. The stretched walls do not buckle, and the narrowing and elongation of the cells in the pull direction reduce the specimen's cross-sectional area considerably. Also, any local failure in a stretched cell wall reduces the overall mechanical integrity of the whole specimen.[22]

MICROSCOPIC PROPERTIES

Light and Polarized Microscopy

A basic approach for visualizing the microstructure of a sample is to obtain a magnified view of its components through microscopic techniques. Various microscopy techniques have been used to study bread and its components. Light microscopy techniques, with a resolution of 200 to 500 nm,[33] have been widely used for observing bakery products. For example, light microscopy has been used to observe the effect of mixing on the structure of dough made from different quality flours.[34] Staining gluten fractions with Uranyl-acetate shows that proper dough development requires a matrix network of protein strands.

Polarized light microscopy has been used to show changes in starch as bread stales. A polarized microscope is used to detect highly ordered material by the use of polarizing prisms or filters which block out all reflective light except that originating from the part of the specimen oriented in a particular direction. Native starch granules are birefringent and show a characteristic Maltese cross pattern under a polarized microscope. Birefringence implies a high degree of molecular orientation within the granule.[35] Upon gelatinization, the crystalline component becomes amorphous and loses birefringence. As bread is aged, the starch begins to retrograde, physically changing from a swollen, gel-like state to a more crystalline state[36] and re-exhibiting the Maltese cross in a polarized microscope. Rao and coworkers[29] used polarized light microscopy to show that starch crystals in stale bread form the discontinuous phase in a continuous amorphous matrix.

Confocal Microscopy

The main drawbacks of light microscopy are limited resolution and the need to obtain thin sections for analysis. A specimen with considerable thickness yields a blurred or diffused image, since the depth of field in conventional microscopy at high power is 2 to 3μ, while the resolving power is 0.2μ.[37] The confocal microscope circumvents this problem by focusing a point light source on a small volume within the specimen, rendering an image of in-focus plane only, with the out of focus parts appearing as black background.[38] The confocal microscope functions by focusing a light source on a x-y plane at a specific depth (z). The illumination is often coherent (laser) light, since its greater intensity allows for additional applications such as observing fluorescence in a specimen. Further details of this technique can be found elsewhere.[39-41] The location and distribution in 3-D of bread components (starch and gluten) was studied using confocal microscopy.[42] The confocal micrographs of fresh and ten day old bread were identical. This technique could not differentiate between the state of the different components (crystalline or amorphous). A combination of techniques would be necessary to obtain a more complete picture of changes that occurred in the bread.

Electron Microscopy

Electron microscopy provides better resolution (15 to 20 nm) and higher magnifications than conventional light microscopy,[43] since the illumination source consists of electrons focused with magnetic lenses rather than photons focused with glass lenses. Scanning electron microscopy (SEM) is used to examine surfaces. Observation of bread by SEM can provide information on gas cell size. However, the use of fixatives and removal of moisture can cause the protein matrix to shrink away from the starch granules, altering the appearance of the bread components.[44,45] Cryo-SEM can circumvent some of these problems since it involves freezing the specimen and examining it at a temperature at which water is not lost and fat is not melted. Structure organization in cereals, changes during heating or baking,[45,46] dough formation, effect of mixing on the gluten network,[47,48] and examination of cake batters[49] are some of the many applications of cryo-SEM in the study of cereal products.[50] Chapter 8 explains the application of microscopy on bread staling.

MAGNETIC RESONANCE IMAGING (MRI)

Nuclear magnetic resonance imaging (MRI) is a spectroscopy technique that can be used to obtain a three-dimensional representation of the systems' molecular properties. Details of the MRI technique have been extensively discussed elsewhere,[51-53] and its application in bread staling studies is reported in Chapter 6.

STRUCTURAL PROPERTIES

DIFFERENTIAL SCANNING CALORIMETRY (DSC)

Differential scanning calorimetry (DSC) monitors changes in physical or chemical properties of a material as a function of temperature by detecting the heat changes associated with such processes. DSC compares the rate of heat flow of a sample to that of an inert material as both are heated or cooled. Aging of bread and starch gels showed development of an endotherm at around 60°C.[54,55] This staling (S) endotherm depicts the development of ordered structure (including crystallization) of amylopectin. This endotherm was found in various kinds of aged starch gels.[56,57] DSC has also been used to monitor the amount of freezable water in starch gel and bread.[58,59] DSC has been used to measure T_g of bread[60] and starch gels[61] by observing a heat capacity change. However, DSC lacks the sensitivity to accurately measure T_g of biopolymers,[1,3] and the glass transition is difficult to observe in DSC since it involves a small baseline shift. In some cases, a rescan of the sample at the same heating rate may confirm its presence.[62] The rescan should involve rapid cooling after the initial scan to prevent recrystallization and therefore, subsequent melting. Utilizing this method, a glass transition would appear in both scans, as shown by Kalichevsky and coworkers,[63] for amylopectin. Nonetheless, explanation of thermal analyses is extremely difficult for multicomponent systems such as bread.[64] Various constituents have different domains with overlapping transitions over a wide temperature range.

DYNAMIC MECHANICAL ANALYSIS (DMA)

Dynamic mechanical analysis (DMA) is a thermomechanical technique that applies dynamic stress at a given frequency to a sample of known geometry. The resulting strain has two components, in phase (elastic) and out of phase (viscous).[65] Food polymers are mostly viscoelastic, with stress and strain being out of phase with respect to each other by a phase angle (δ). tan δ is defined as the ratio of E''/E'. E'' is the loss modulus which reflects the energy dissipated as the material is deformed, and E' is the storage modulus which indicates the storage of energy.[66,67]

At a given temperature, several domains of a semicrystalline polymer differ in their degrees of segmental mobility. DMA can differentiate between these domains by monitoring the E′, E″, and tan δ. A thermomechanical transition region (either crystal-to-melt or glassy-to-rubbery) is evidenced in DMA by a drop in E′, a peak in E″, and a peak in tan δ. These parameters have been monitored for standard white bread[59] and meal, ready-to-eat (MRE) bread.[68] The observed tan δ curve can span over a wide range of temperature (50 to 100°C), and its breadth and midpoint temperature can be increased with decreasing moisture and increasing storage time. Thermomechanical analysis (TMA) has also been applied to measure thermomechanical transition of bread.[69] Similar trends were observed in gluten[70,71] and starch gels.[72] It is clear that in bread polymers a distribution of transition temperatures occurs which can only be described using a distribution function rather than an onset or midpoint temperature.[59,73]

Thermomechanical transitions of standard white bread, as observed by DMA tan δ (T), changed with storage time (Figure 2).[4] After curve decomposition, the main transition (a Gaussian curve, G in Figure 2) represents ice melting, since it was found to correlate well with DSC freezable water content in the bread.[59] Over staling, G peak decreased in amplitude as freezable water decreased with time. A second A curve (asymmetric double sigmoidal, Figure 2) seems to be associated with the solid domains. The A peak also decreased in amplitude and broadened with storage time. Similar transitions were observed for MRE bread,[68] low moisture bread,[74] and starch gels.[72]

The DMA loss modulus (E′) can also be used to follow changes occurring in the sample with increasing temperature. Peleg[75] has been a pioneer in this area with a model equation based on Fermi's distribution function to describe the change of the E′ with temperature within a range of moisture contents for different food polymers. For example, a fitted curve and its corresponding experimental E′ (T) values are depicted in Figure 3 for white bread.[76] From each fit, two parameters were obtained: "a" the steepness or the slope of the curve, and "T_c", the temperature at the inflection point in the curve. A plot of T_c against moisture content for different baked goods is depicted in Figure 4. The critical temperature decreased sharply for standard white bread and gluten films between 0 and 20% moisture, while much less so for MRE bread, due to added glycerol.[68] The effect of moisture on E′ can be represented in a 3-D contour plot[75] which was used for various baked products (Figure 5). Given two conditions such as moisture and temperature, the third parameter, in this case stiffness, may be predicted within the range of the experiments. Table 1 includes the "T_c" and "a" equations for various bakery items as a comparison. A higher value for "a" reflects a more gradual drop in E′ (T) and, therefore, a more extended (larger temperature range) transition region, while an increased or decreased "T_c" value would result from the transition region moving to higher or lower temperature, respectively.

MOLECULAR PROPERTIES

The changes detected in bread during storage are the tangible manifestation of phenomena that take place in the product at both structural and molecular level.

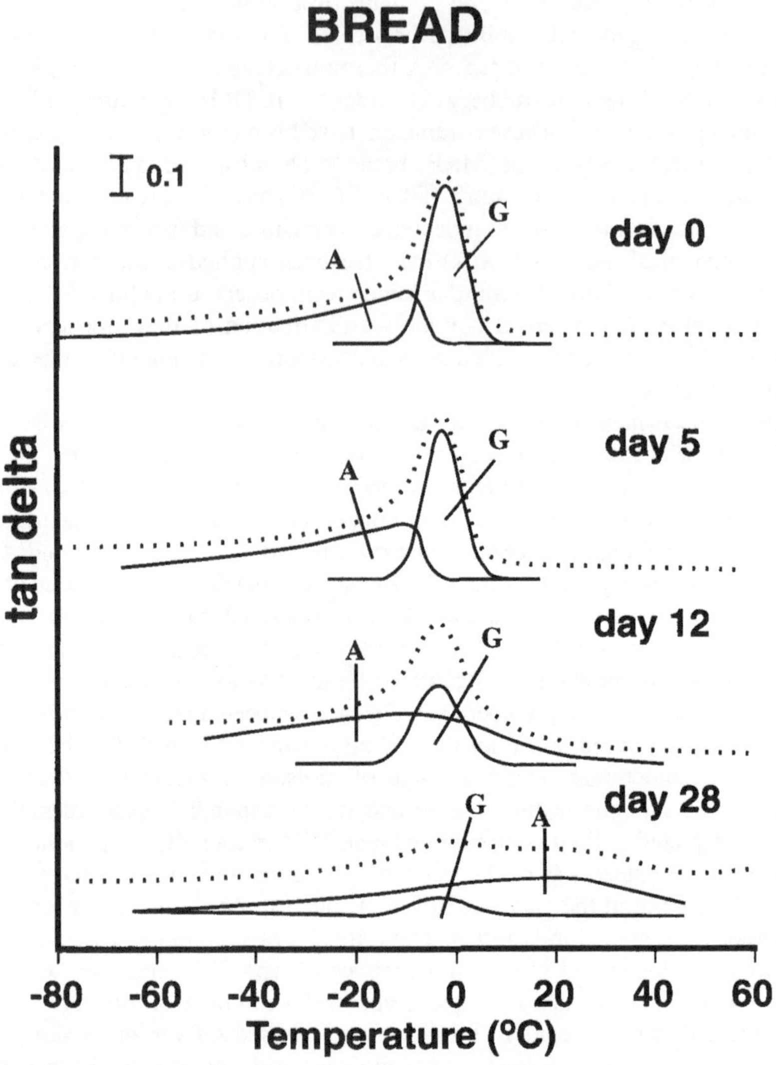

FIGURE 2 DMA tan δ (T) deconvolution results of white bread stored for 28 days at ambient temperature. The asymmetric double Sigmoidal curve (A) and Gaussian curve (G) are designated.[4]

Molecular properties deal with events in microseconds/milliseconds and can be analyzed with molecular spectroscopic methods such as nuclear magnetic resonance (NMR) and electron spin resonance (covered in detail in Chapter 7).

Nuclear Magnetic Resonance

NMR has been applied to investigate molecular changes in breads and bread polymers.[5,77-81] NMR is based on the ability of nuclei with magnetic dipole moments to absorb electromagnetic energy. Nuclei absorb characteristic frequencies of electromagnetic radiation depending on the strength of the magnetic field and the chemical

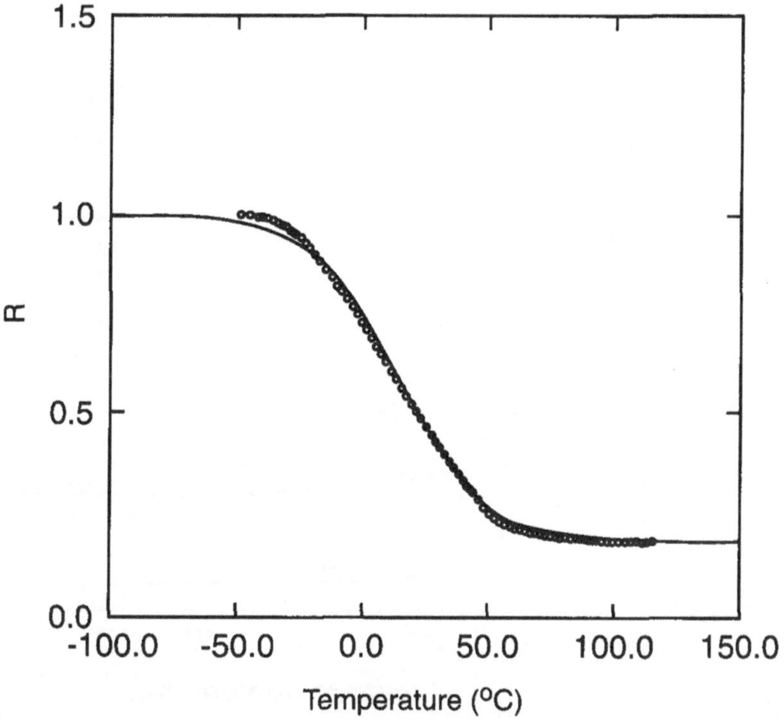

FIGURE 3 Normalized E′ (T) curve (R) obtained from DMA analysis of a 6.7% moisture bread sample (symbols) with the fitted results (solid line) from the modified Fermi equation.

and magnetic environment. Details on the various NMR methodologies can be found elsewhere.[1,4,82,83]

Proton NMR

Proton (^1H) is the most abundant NMR-detectable species, allowing for relatively easy acquisition and strong signal. However, the ^1H relaxation process is perturbed by other phenomenon, including cross-relaxation and chemical exchange.[84-86] Nonetheless, wide line ^1H NMR[87] can be used to ascertain the physical state of the protons in a system, i.e., solid like (wide) and more liquid like (narrow) components.[88,89] Protons associated with the solid state can be studied by solid-state proton NMR, such as water sorption in starch and gluten[58,89] and cross-relaxation (CR) NMR spectroscopy.[88,90]

(CR) NMR is used to determine information on the solid component relaxation via the observable liquid spin system.[90] A sample is irradiated with a radio-frequency pulse, which is off-resonance from the liquid signal. Due to the dipolar coupling between the liquid and solid protons, the amplitude of the liquid spectrum will change with the offset frequency, and a Z-spectra is obtained.[88,90] Wu and Eads[91] found an increase in Z-spectral line shape, area, and width for waxy starch gels during aging which was dependent on concentration and storage.[88]

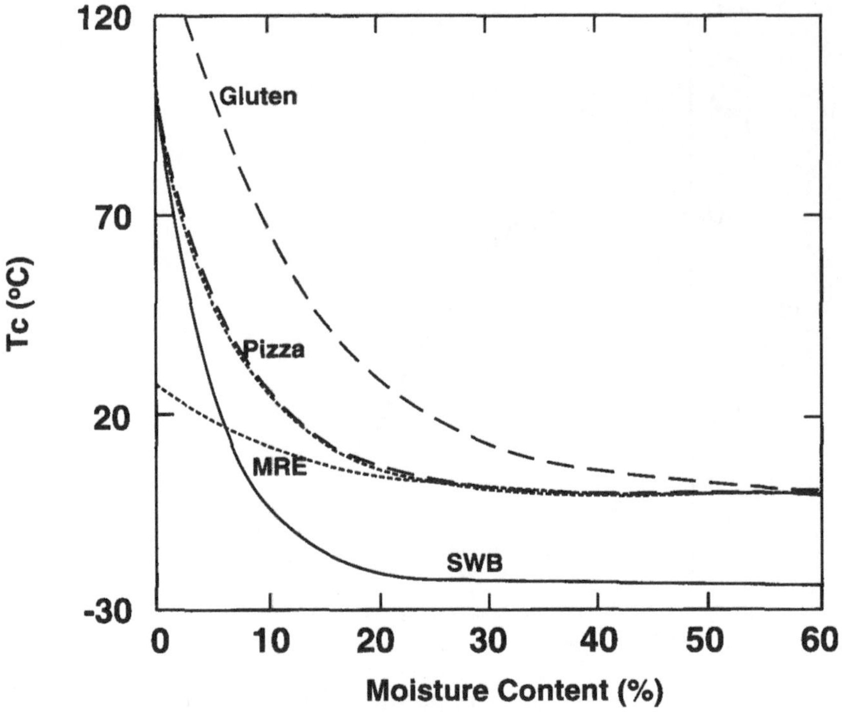

FIGURE 4 The change of "T_c" (the midpoint temperature of the fitted E′ (T) curve) in standard white bread (SWB), gluten, pizza (microwaved and conventionally cooked), and meal ready to eat (MRE) samples adjusted to variable moisture contents.

These Z-spectra can be further analyzed and deconvoluted into their constituent components.[90-92] The best fit for wheat starch gels (Figure 6 dashed lines) required a combination of Lorentzian (mobile) and Gaussian (immobile) functions (Figure 6 solid lines). In contrast, a freshly gelatinized waxy maize starch (amylopectin) yielded a single Lorentzian function,[91] suggesting that recrystallized amylose contributed to the rigid solid component in the freshly gelatinized wheat starch. Development of the rigid component in wheat starch gel (60% moisture) during aging is shown in Figure 7. The area of the Gaussian component increased in 0 to 8 days, indicating starch crystallization, whereas that of the Lorentzian component decreased within this period.[91]

Deutrium NMR

Deuterium (^2H) NMR can be used to study motion of water in solid polymers[93] such as starch and gluten[58,71,93,94] and in complex matrix, such as bread.[4] Solid state ^2H NMR of aging bread indicated that all the deuterons were in fast exchange conditions and underwent isotropic reorientation in the sample; in addition their mobility did not change during storage.[4] The high mobility of the water in bread was confirmed by ^{17}O NMR,[79] the only NMR technique specific for water.[82] Kim-Shin and

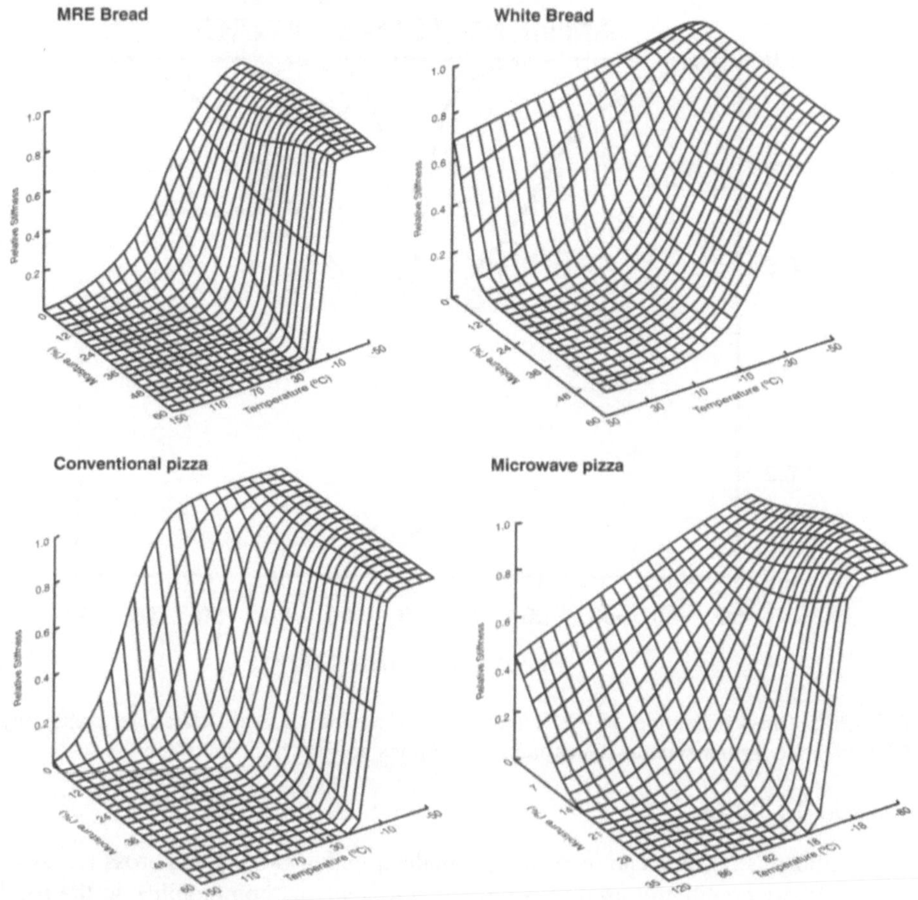

FIGURE 5 Three-dimensional contour plots of the relative stiffness (calculated from the E′ (T) curve of DMA)-temperature-moisture relationship of MRE bread, white bread, conventionally heated pizza, and microwaved pizza.

TABLE 1
T_c and a Equations for Bakery Items (M is moisture content)

Item	Temperature at inflection "Tc"	Steepness of curve "a"
Standard white bread	Tc = 122.6 * exp (–0.185*M)	A = 73.0 * exp (–0.244*M)
Meal ready to eat bread	Tc = 28.12 * exp (–0.082*M)	A = 26.11 * exp (–0.043*M)
Conventionally heated pizza	Tc = 90.56 * exp (–0.122*M)	A = 16.00 * exp (–0.027*M)
Microwaved pizza	Tc = 96.35 * exp (–0.13*M)	A = 97.63 * exp (–0.098*M)

coworkers[79] also indicated that the amount of ^{17}O NMR–detectable water, and its mobility, slightly decreased during bread storage, but no correlation between NMR data and recrystallized amylopectin was observed.

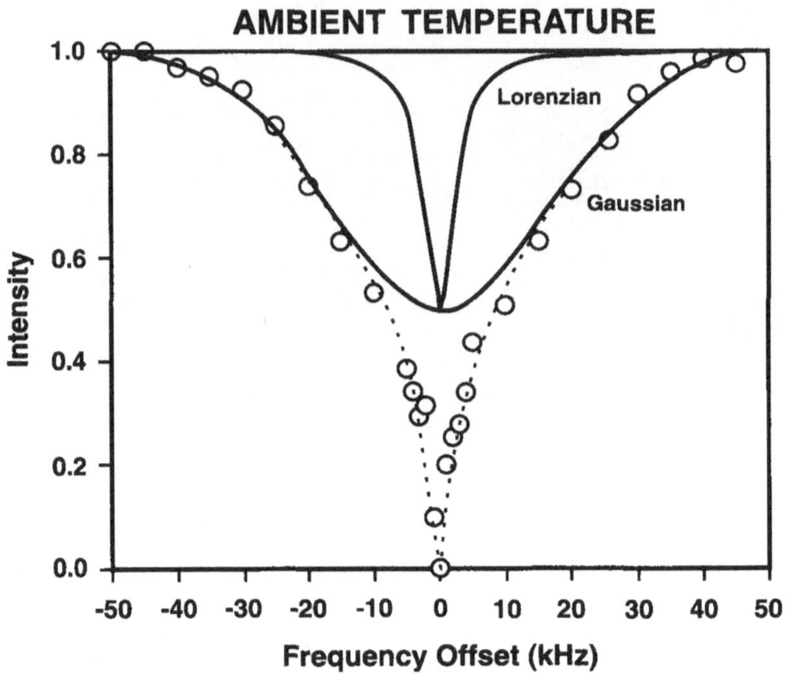

FIGURE 6 Cross relaxation ^1H NMR spectra (open circles) of wheat starch gel containing 60% moisture. Lorentzian and Gaussian contributions are shown in solid line.

CARBON-13 NMR

Two carbon-13 NMR techniques, magic angle spinning (MAS) and cross polarization (CP), are commonly used in conjunction to study carbon mobility in the solid state. If a sample is spun rapidly at an angle of 54.74° to the applied magnetic field, the dipolar interactions are reduced. At this magic angle, the chemical shift tensor (σ) is equal to the isotropic chemical shift (σ_i) which is found in solutions when dipolar interactions are not seen. MAS, therefore, reduces the broad chemical shift powder pattern to the isotropic sharp pattern found in liquids.[93] For nuclei with low natural abundance (such as ^{13}C), the average of many scans needs to be accumulated to obtain a spectrum when using Fourier transform pulsed techniques. The repetition rate is dictated by T_1 (spin-lattice relaxation time), and for ^{13}C, the T_1 values are very long in solids and, therefore, the accumulation of the data required would take a prohibitively long time. Consequently, CP is used, in which the polarization of the abundant ^1H nuclear spin with short T_1 is transferred to that of the ^{13}C nuclei. After CP, the repetition rate is determined by the short ^1H T_1s.[93,95] Therefore, by utilizing CP-MAS NMR, resolution and experimental time frame have both improved.

^{13}C CP MAS NMR can be used to monitor the development of rigid components in starch.[96,97] This method was applied to the study of different crystalline structures (A and B pattern) of starch[97] and the effect of hydration.[63] Type A leads to a triplet pattern, whereas Type B leads to a doublet pattern of the carbon 1 peak.[96,97] The

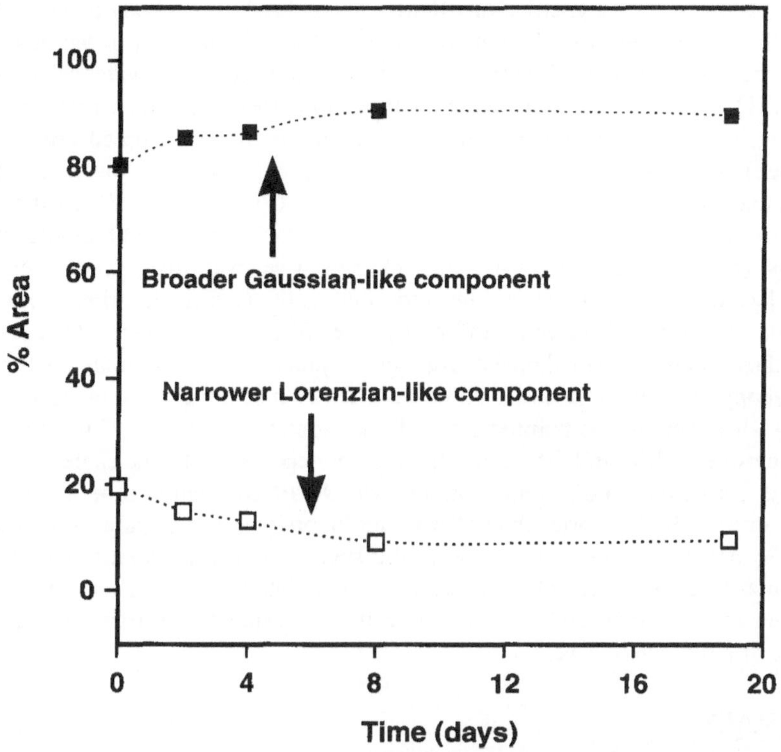

FIGURE 7 Changes in narrow and broad components of wheat starch gels (60% moisture) at 25°C.

mobility of the amorphous and crystalline solid components of wheat starch can be monitored using the rotating frame relaxation time ($T_{1\rho}$) experiment.[5,58,98]

Pulsed NMR

Molecular motions can be studied in various pulsed NMR experiments to obtain relaxation times.[99] Longitudinal relaxation, T_1, is defined as the time constant that characterizes the rate at which the z vector component of magnetization returns to its equilibrium value. Transverse relaxation, T_2, is defined as the time constant that characterizes the rate of decay of the magnetization in the x-y plane. Upon storage or cooling, T_1 and T_2 of bread and bread components indicated a decrease in ^{1}H NMR mobility[4,74,80,81,87,100,101] and ^{2}H NMR mobility,[77] with increasing storage time and decreasing temperature. Interpretation of proton NMR is complicated by proton exchange and cross relaxation.[84-86] Hence, contribution of specific components cannot be easily characterized.

Pulsed field gradient (PFG) NMR is a very specific method of examining water movement and the degree of water-polymer association without destroying the integrity of the sample.[102] In this method, the two field gradient pulses allow the nuclei to be positionally identified, and translational movement can be determined by refocusing of the magnetization after the second radio frequency pulse.[103] Therefore, PFG-NMR measures Brownian motion of water molecules in the absence of concentration gradient, leading to a noninvasive determination of water self-diffusion coefficient. Observed water diffusion coefficient in bread normally corresponds with changes in moisture content during storage. For instance, in bread crumb (crust included) stored at 25°C in tri-laminated pouches, the water diffusion coefficient was found to decrease from 0.67×10^{-6} cm^2/s to 0.34×10^{-6} cm^2/s as the crumb moisture content decreased from 39 to 31%. Similarly, breadcrumb stored without its crust (no moisture change) had little change in water diffusion coefficient.

Selected NMR methods can be important tools used in describing molecular motions of starch and water in different phases or physical states.[4,58,89] The results from these methods can then be used in conjunction with thermal analysis and microscopy techniques to better understand the changes occurring in starch during staling. However, as was pointed out by Dickinson and coworkers,[104] the correlation times between NMR and DMA or DSC are not expected to be the same, since each method is sensitive to different motions, with NMR correlation time being several orders of magnitude shorter than DMA. Additionally, what appears rigid at a DSC or DMA time scale may contain molecular species with considerable mobility at a molecular level. For example, unfreezable water (DSC), which has been described as water in a vetrified and viscous phase in waxy starch was found to be highly mobile at a molecular level.[89]

SUMMARY

The combination of various techniques can be a powerful approach to the study of physiochemical changes of complex phenomena. Choosing an appropriate method becomes a significant factor in interpretation of results in time-dependent processes such as the glass transition, recrystallization, and drying. Since the glass transition usually describes segmental mobility, extending its meaning to molecular mobility requires careful investigation using molecular spectroscopic techniques. It is important to define whether the mobility is molecular, segmental, or on a larger scale. The results of measuring these kinetic events are dependent on experimental parameters such as heating or cooling rates and sample characteristics. Methods that have been described in this chapter determine local molecular rotational mobility, structural domain properties, microscopic nature, and bulk rheological properties. Careful selection of techniques is a critical step. Scientists need to approach this complex problem with well-planned strategies so experimental data can serve as building blocks for a fundamental understanding of molecular structure and function relationships in bread.

REFERENCES

1. Chinachoti, P., Probing molecular and structural thermal events in cereal-based products, *Thermochimica Acta*, 246, 357, 1994.
2. Chinachoti, P., Water migration and food storage stability, in *Food Storage Stability*, Taub, I.A. and Singh, R.P., Eds., CRC Press LLC, Boca Raton, 1998, 245.

3. Chinachoti, P., Characterization of thermomechanical properties in starch and cereal products, *J. Thermal Anal.*, 47, 195, 1996.
4. Chinachoti, P., NMR dynamic properties of water in relation to thermal characteristics in bread, in *The Properties of Water in Foods ISOPOW 6,* Reid, D.S., Ed., Blackie Academic & Professional, London, 1998, 139.
5. Vodovotz, Y. and Chinachoti, P., Probing molecular motions of low moisture starch gels by carbon-13 nuclear magnetic resonance, in *Applications of Magnetic Resonance to Food Science*, Belton, P.S., Hills, B.P., and Web, G., Eds., Royal Society of Chemistry, Cambridge, 1999, 185.
6. Peleg, M., Roy, I., Campanella, O.H., and Normand, M.D., Mathematical characterization of the compressive stress-strain relationships of spongy baked goods, *J. Food Sci.*, 54, 947, 1989.
7. Spies, R., Application of rheology in the bread industry, in *Dough Rheology and Baked Product Texture*, Faridi, H. and Faubion, J.M., Eds.,Van Nostrand and Reinhold, New York, 1990, 343.
8. Axford, D.W.E., Colwell, K.W., Conford, S.J., and Elton, G.A.H., Effect of loaf specific volume on the rate and extent of staling in bread, *J. Sci. Food Agric.*, 19, 95, 1968.
9. Pomeranz, Y. and Shellenberger, J.M., Sensory attributes and bread staling, in *Bread Science and Technology,* Pomeranz, Y. and Shellenberger, J.M., Eds., AVI Publishing, Westport, CT, 1971.
10. AACC Method 44-18 and Method 74-09, Bread firmness by universal testing machine, *American Association of Cereal Chemists*, St. Paul, Minnesota, USA. 1983.
11. Ponte Jr., J.G. and Ovadia, D.Z., Instrumental methods, in *Baked Goods Freshness: Technology, Evaluation, and Inhibition of Staling*, Hebeda, R.E. and Zobel, H.F., Eds., Marcel Decker, New York, 1996, 151.
12. Lasztity, R., Rheological studies on bread at the technical University of Budapest, *J. Texture Stud.*, 11, 81, 1980.
13. Baker, A.E. and Ponte J.G., Jr., Measurement of bread firmness with the universal testing machine, *Cereal Foods World*, 32, 491, 1987.
14. Avrami, M., Kinetics of phase change I, *J. Chem. Phys.*, 7, 1103, 1939.
15. Avrami, M., Kinetics of phase change II, *J. Chem. Phys.,* 8, 212, 1940.
16. Avrami, M., Kinetics of phase change III, *J. Chem. Phys.,* 9, 177, 1941.
17. Cornford, S.J., Axford, D.W.E., and Elton, G.A.H., The elastic modulus of bread crumb in linear expression in relation to staling, *Cereal Chem.*, 41, 216, 1964.
18. McIver, R.G., Axford, D.W.E., Colwell, K.H., and Elton, G.A.H., Kinetic study of the retrogradation of gelatinized starch, *J. Sci. Food Agric.,* 19, 560, 1968.
19. Kim, S.K. and D'Appolonia, B.L., Bread staling studies. I. Effect of protein content on staling rate and bread crumb pasting properties, *Cereal Chem.*, 54, 207, 1977.
20. Willhoft, E.M.A., Bread staling. II. Theoretical study, *J. Sci. Food Agric.,* 22, 180, 1971.
21. Ashby, M.L., The mechanical properties of cellular solids, *Metall. Trans.,* 14A, 1755, 1983.
22. Peleg, M., Review: mechanical properties of dry cellular solid foods, *Food Sci. Tech. Int.,* 3, 227, 1997.
23. Attenborrow, G.E., Goodband, R.M., Taylor, L.J., and Lillford P.J., Structure, mechanics and texture of a food sponge, *J. Cereal Sci.*, 9, 61 1989.
24. Swyngedau, S., Nussinovitch, A., Roy, I., Peleg, M., and Huang, V., Comparison of four models for the compressibility of breads and plastic foams, *J. Food Sci.,* 56, 756, 1991.

25. Kou, Y. and Chinachoti, P., Structural damage in bread staling as detected by recoverable work and stress-strain model analysis, *Cereal Foods World*, 36, 888, 1991.
26. Hibberd, G.E. and Parker, N.S., Measurements of the compression properties of bread crumb, *J. Texture Stud.*, 16, 97, 1985.
27. Nussinovitch A., Steffens, M., and Chinachoti, P., Elastic properties of breadcrumb, *Cereal Chem.*, 69, 678, 1992.
28. Nussinovitch, A., Steffens, M., Chinachoti, P., and Peleg, M., Effect of strain level and storage time on the recoverable work of compressed breadcrumb, *J. Texture Stud.*, 23, 13, 1992.
29. Rao, P.A., Nussinovitch, A., and Chinachoti, P., Effects of selected surfactants on amylopectin recrystallization and recoverability of breadcrumb during storage, *Cereal Chem.*, 69, 613, 1992.
30. Platt, W. and Kratz, P.D., Measuring and recording some characteristics of test sponge cakes, *Cereal Chem.*, 10, 73, 1933.
31. Bourne, M.C., *Food Texture and Viscosity: Concept and Measurement*, Academic Press, New York, 1982.
32. Chen, P. Whitney, L.F., and Peleg, M., Some tensile characteristics of bread crumb, *J. Texture Stud.*, 25, 299, 1994.
33. Aguillera, J.M. and Stanley, D.W., *Microstructural Principles of Food Processing and Engineering*, Elsevier, New York, 1990.
34. Bechtel, D.B ., Poranz, Y., and de Francisco, A., Breadmaking studied by light and transmission electron microscopy, *Cereal Chem.*, 55, 392, 1978.
35. Varriano-Marston, E., Polarization microscopy: applications in cereal science, in *New Frontiers in Food Microstructure*, Bechtel, D.B., Ed., American Association of Cereal Chemists, St. Paul, 1983, 71.
36. Krog, N., Olesen, S.K., Ternaes, H., and Joensson, T., Retrogradation of the starch fraction in wheat bread, *Cereal Foods World*, 34, 281, 1989.
37. Lemasters, J.J., Chacon, E., Zahrebelski, G., Reece, J.M., and Nieminen, A-L., Laser scanning confocal microscopy of living cells, in *Optical Microscopy: Emerging Methods and Applications*, Herman, B. and Lemasters, J.J., Eds., Academic Press, San Diego, CA, 1993.
38. Cavanagh, H.D., Petroll, W.M., and Jester, J.V., Confocal microscopy: uses in measurement of cellular structure and function, *Prog. Retinal Eye Res.*, 14, 527, 1995.
39. Blonk, J.C.G. and van Aalst, H., Confocal scanning light microscopy in food research, *Food Res. Int.*, 26, 297, 1993.
40. Vodovotz, Y., Vittadini, E., Coupland, J., Chinachoti, P., and McClements, D.J., Bridging the gap: the use of confocal microscopy in food research, *Food Technol.*, 50, 74, 1996.
41. Vodovotz, Y. and Tsoubelli, M., Food microstructure, in *Wiley Encyclopedia of Food Science*, 2nd ed., Francis, F.J., Ed., 2, 939, 1999.
42. Vodovotz Y. and Chinachoti, C., Confocal microscopy of bread, in *New Techniques in the Analysis of Foods*, Tunick, M.H., Palumbo, S.A., and Fratamico, P.M., Eds., Plenum Press, New York, 1998, 9.
43. Postek, M.T., Howard, K.S., Johnson, A.H., and McMichael, K.L., *Scanning Electron Microscopy-a Student's Handbook*, Ladd Research Industries, Williston, VT, 1980.
44. Chabot, J.F., Hood, L.F., and Liboff, M., Effect of scanning electron microscopy preparation methods on the ultrastructure of white bread, *Cereal Chem.*, 56, 462, 1979.
45. Freeman, T.P. and Shelton, D.R., Microstructure of wheat starch: from kernel to bread, *Food Technol.*, 3, 162, 1991.

46. McDonough, C.M., Seetharaman, K., Waniska, R.D., and Rooney, L.W., Microstructure changes in wheat flour tortillas during baking, *J. Food Sci.,* 61, 995, 1996.
47. Evans, L.G., Pearson, A.M., and Hooper, G.R., Scanning electron microscopy of flour-water doughs treated with oxidizing and reducing agents, *Scanning Electron Microsc. III,* 583, 1981.
48. Autio, K. and Laurikainen, T., Relationships between flour/dough microstructure and dough handling and baking properties, *Trends Food Sci. Technol.,* 8, 181, 1997.
49. Hsieh, S.I., Davis, E.A., and Gordon, J.,Cryofixation freeze-etch of cake batters, cereals and cereal products, electron microscopy, *Cereal Foods World,* 26, 562, 1981.
50. Yiu, S.H., Food microscopy and the nutritional quality of cereal foods, *Food Struct.,* 12, 123, 1993.
51. McCarthy, M.J., *Magnetic Resonance Imaging in Foods,* Chapman & Hall, New York, 1994.
52. Schmidt, S.J., Sun, X., and Litchfield, J.B., Applications of magnetic resonance imaging in food science, *Crit. Rev. Food Sci. Nutr.,* 36, 357, 1996.
53. Belton, P.S., Hills, B.P., and Webb, G.A., *Advances in Magnetic Resonance in Food Science,* Royal Society of Chemistry, U.K., 1999.
54. Russell, P.L., A kinetic study of bread staling by differential scanning calorimetry and compressibility measurements. The effect of different grists, *J. Cereal Sci.,* 1, 285, 1983.
55. Fessas, D. and Schiraldi, A., Texture and staling of wheat bread crumb: effects of water extractable proteins and 'pentosans', *Thermochimica Acta,* 323, 17, 1998.
56. Ring, S.G., Colonna, P., I'Anson, K.J., Kalichevsky, M.T., Miles, M.J., Morris, V.J., and Orford, P.D., The gelation and crystallization of amylopectin, *Carbohydr. Res.,* 162, 277, 1987.
57. Miles, M.J., Morris, V.J., Orford, P.D., and Ring, S.G., The roles of amylose and amylopectin in the gelation and retrogradation of starch, *Carbohydr. Res.,* 135, 271, 1985.
58. Li, S., Dickinson, C., and Chinachoti, P., Proton relaxation of starch, gluten and their mixture by solid-state nuclear magnetic resonance, *Cereal Chem.,* 73, 736, 1996.
59. Vodovotz, Y., Hallberg, L., and Chinachoti, P., Effect of aging and drying on thermomechanical properties of white bread as characterized by dynamic mechanical analysis (DMA) and differential scanning calorimetry (DSC), *Cereal Chem.,* 73, 264, 1996.
60. Laine, M.J.K. and Roos,Y., Water plasticization and recrystallization of starch in relation to glass transition, *ISOPOW Practicum II,* 1994
61. Zeleznak, K.J. and Hoseney, R.C., The glass transition in starch, *Cereal Chem.,* 64, 121, 1987.
62. Yost, D.A. and Hoseney, R.C., Annealing and the glass transition of starch, *Starke,* 38, 289, 1986.
63. Kalichevsky, M.T., Jaroszkiewicz, E.M., Ablett, S., Blanshard, J.M.V., and Lillford, P.J., The glass transition of amylopectin measured by DSC, DMTA, and NMR, *Carbohydr. Polym.,* 18, 77, 1992.
64. Biliaderis, C.G. and Galloway, G., Crystallization behavior of amylose-v complexes: structure-property relationships, *Carbohydr. Res.,* 189, 31, 1989.
65. Wendlandt, W.W. and Gallagher, P.K., Instrumentation, in *Thermal Characterization of Polymeric Materials,* Turi, E.A., Ed., Academic Press, New York, 1981, 3.
66. Hamann, D.D., Purkayastha, S., and Lanier, T.C., Applications of thermal scanning rheology to the study of food gels, in *Thermal Analysis of Foods,* Harwalkar, V.R. and Ma, C.Y., Eds., Elsevier, New York, 1990, 306.
67. Murayana, D., *Dynamic Mechanical Analysis of Polymeric Material,* Elsevier, Amsterdam, 1978.

68. Hallberg, L.M. and Chinachoti, P., Dynamic mechanical analysis for glass transitions in long shelf-life bread, *J. Food Sci.*, 57, 1, 1992.
69. LeMeste, M., Huang, V.T., Panama, J., Anderson, G., and Lentz, R., Glass transition of bread, *Cereal Foods World*, 37, 264,1992.
70. Cherian, G. and Chinachoti, P., ^{2}H and ^{17}O nuclear magnetic resonance study of water in gluten in the glassy and rubbery state, *Cereal Chem.*, 73, 618, 1996.
71. Cherian, G. and Chinachoti, P., Action of oxidants on water sorption, ^{2}H nuclear magnetic resonance mobility, and glass transition behavior of gluten, *Cereal Chem.*, 74, 312, 1997.
72. Vodovotz, Y. and Chinachoti, P., Glassy-rubbery transition and recrystallization during aging of wheat starch gels, *J. Agric. Food Chem.*, 46, 446, 1998.
73. Peleg, M., A note on the tan δ (T) peak as a glass transition indicator in biosolids, *Rheologica Acta,* 34, 215, 1995.
74. Roudaut, G., Maglione, M., van Dusschoten, D., and Le Meste M., Molecular mobility in glassy bread: a multispectroscopy approach, *Cereal Chem.*, 76, 70, 1999.
75. Peleg, M., Mapping the stiffness-temperature-moisture relationship of solid biomaterials at and around their glass transition, *Rheologica Acta,* 32, 575, 1993.
76. Vodovotz, Y., Aging of Starch and Bread as Studied by DSC, DMA, NMR and Confocal Microscopy, Ph.D. Thesis, University of Massachusetts, 1996.
77. Leung, H.K., Magnuson J.A., and Bruinsma, B.L., Water binding of wheat flour doughs and breads as studied by deuteron relaxation, *J. Food Sci.*, 48, 95, 1983.
78. Wynne-Jones, S. and Blanshard J.M.V., Hydration studies of wheat starch, amylopectin, amlose gels and bread by proton magnetic resonance, *Carbohydr. Polym.*, 6, 289, 1986.
79. Kim-Shin, M-S., Mari, F., Rao, P.A., Stengle, T.R., and Chinachoti, P., ^{17}O Nuclear magnetic resonance studies of water mobility during bread staling, *J. Agric. Food Chem.*, 39, 1915, 1991.
80. Ruan, R., Almaer, S., Huang, V.T., Perkins, P., Chen, P., and Fulcher, R.G., Relationship between firming and water mobility in starch-based food systems during storage, *Cereal Chem.*, 73, 328, 1996.
81. Chen, P.L., Long, Z., Ruan, R., and Labuza, T.P., Nuclear magnetic resonance studies of water mobility in bread during storage, *Lebensm. Wiss Technol.*, 30, 176, 1997.
82. Richardson, S.J. and Steinberg, M.P., Applications of nuclear magnetic resonance, in *Water Activity Theory and Applications to Food*, Rockland, L.B. and Beuchat, L.R., Eds., Marcel Dekker, New York, 1987.
83. Belton, P.S., Can nuclear magnetic resonance give useful information about the state of water in food stuffs? *Comments Agric. Food Chem.*, 2, 179, 1990.
84. Halle, B. and Wennestrom, H., Interpretation of magnetic resonance data from water nuclei in heterogeneous systems, *J. Phys. Chem.*, 75, 1928, 1981.
85. Schmidt, S.J. and Lai, H.M., Use of NMR and MRI to study water relations in foods, in *Water Relationships in Foods*, Levine, H. and Slade, L., Eds., Plenum Press, New York, 1991, 405.
86. Colquhoun, I.J. and Goodfellow, B.J., Nuclear magnetic resonance spectroscopy, in *Spectroscopic Techniques for Food Analysis*, Wilson, R.H., Ed., VCH Publishers, New York, 1994, 87.
87. Tanner, S.F., Hills, B.P., and Parker, R., Interactions of sorbed water with starch studied using proton nuclear magnetic resonance spectroscopy, *J. Chem. Soc. Faraday Trans.*, 87, 2613, 1991.
88. Wu, J.Y., Bryant, R.G., and Eads, T.M., Detection of solidlike components in starch using cross-relaxation and Fourier transform wide-line ^{1}H NMR methods, *J. Agric. Food Chem.*, 40, 449, 1992.

89. Li, S., Dickinson, L.C., and Chinachoti, P., Mobility of "unfreezable" and "freezable" water in waxy corn starch by ^2H and ^1H NMR, *J. Agric. Food Chem.*, 46, 62, 1998.
90. Grad, J. and Bryant, R.G., Nuclear magnetic cross-relaxation spectroscopy, *J. Magn. Resonance*, 90, 1, 1990.
91. Wu, J.Y. and Eads, T.M., Evolution of polymer mobility during aging of gelatinized waxy maize starch: a magnetization transfer ^1H NMR study, *Carbohydr. Polym.*, 20, 51, 1993.
92. Henkelman, R.M., Huang, X., Xiang, Q.S., Stanisz, G.J., Swanson, S.D., and Bronskill, M.J., Quantitative interpretation of magnetization transfer, *Magn. Resonance Med.*, 29, 759, 1993.
93. Tonelli, A.E., *NMR Spectroscopy and Polymer Microstructure: The Conformational Connection,* VCH Publishers, New York, 1989.
94. Yakubu, P.I., Baianu, I.C., and Orr, P.H., Deuterium nuclear magnetic resonance studies of potato starch hydration, in *Water Relationships in Foods*, Levine, H. and Slade, L., Eds., Plenum Press, New York, 1991, 585.
95. Pines, A., Gibby, M.G., and Waugh, J.S., Proton-enhanced NMR of dilute spins in solids, *J. Chem. Phys.*, 59, 569, 1973.
96. Blanshard, J.M.V., Jaroszkiewicz, E.M., and Gidley, M.J., The structure and behavior of the starch granule as studied by NMR, in *NMR Applications in Biopolymers. ACS Symposium Series*, Finley, J.W., Schmidt, S.J., and Serianni, A.S., Eds., Plenum Press, New York, 1990, 155.
97. Veregin, R.P., Fyfe, C.A., Marchessault, R.H., and Taylor, M.G., Characterization of the crystalline A and B starch polymorphs and investigation of starch crystallization by high-Resolution ^{13}C CP/MAS NMR, *Macromolecules,* 19, 1030, 1986.
98. Morgan, K.R., Furneaux, R.H., and Stanley, R.A., Observation by solid-state ^{13}C CP MAS NMR spectroscopy of the transformation of wheat starch associated with the making of bread, *Carbohydr. Res.*, 235, 15, 1992.
99. Derome, A.E., *Modern NMR techniques for Chemistry Research,* Pergamon Press, Oxford, 1987.
100. Lechert, H. and Hennig, H.J., NMR investigations on the behavior of water in starches, in *Magnetic Resonance in Colloid and Interface Science*, Resing, H.A. and Wade, C.G., Eds., American Chemical Society, Washington, D.C., 1976, 328.
101. Belton, P.S., Hills, B.P., and Raimbaud, E.R., The effect of morphology and exchange on proton NMR relaxation in agarose gels, *Mol. Phys.*, 63, 825, 1988.
102. Umbach, S.L., Davis, E.A., Gordon, J., and Callaghan, P.T., Water self-diffusion coefficients and dielectric properties determined for starch-gluten-water mixtures heated by microwave and by conventional methods, *Cereal Chem.,* 69, 637, 1992.
103. Blum, F.D., Pulsed-gradient spin echo nuclear magnetic resonance spectroscopy, *Spectroscopy,* 1, 32, 1986.
104. Dickinson L.C., Morganelli, P., Chu, C.W., Petrovic, Z., Macknight, J., and Chien, J.C.W., Molecular motions in model network polymers, *Macromolecules*, 21, 338, 1988.

6 Nuclear Magnetic Resonance Techniques

R. Roger Ruan and Paul L. Chen

CONTENTS

INTRODUCTION

Bread staling is referred to as deterioration in structure, texture, and flavor characteristics during storage — the crust of staled bread becomes soft and leathery, its crumb becomes firm, and its fresh baked flavor is lost. Firming of bread has been the most important parameter in the study of bread staling. As discussed elsewhere in this book, several theories have been proposed to explain why bread firms during storage. It seems that there is no one mechanism that can fully explain the staling phenomenon. No matter what mechanism, be it loss or redistribution of water, change in water property, starch recrystallization, change in gluten protein networks, or interactions between gluten protein and starch granules, bread firming will involve changes in molecular mobility of the bread system, which can be assessed by nuclear magnetic resonance (NMR) techniques.[1-4]

Low field proton NMR has been the choice for study of bread staling because of its ability to rapidly determine the mobility of protons associated with different molecules. It is generally accepted that mobile (liquid phase) protons and immobile (solid phase) protons coexist in a given equilibrium state of food systems. Any physiochemical changes may result in a new equilibrium state of proton mobility. With NMR techniques, one can follow the change in molecular mobility of bread system as it stales.

Seow and Teo[1] used a simple NMR technique to determine the liquid phase and solid phase protons in bread during storage. They found that the decrease in liquid

0-8493-8790-6/01/$0.00+$.50

phase and increase in solid phase protons were highly correlated with bread firming. This is also true for corn starch gel, which suggests that starch retrogradation is solely responsible for the firming of starch gel. The role of starch retrogradation in bread staling is also confirmed by another NMR study by Morgan et al.,[3] who used starch bread for their experiments. Chen et al.[4] used a three-exponential model to analyze the relaxation data for bread and demonstrated interesting changes in proton intensity and mobility of bread that firmed during storage.

In this chapter, the basic NMR and MRI (magnetic resonance imaging) methodologies will be covered; use of NMR and MRI to study moisture migration, proton mobility, and glass transition in bread and other baked products will be discussed.

NMR METHODOLOGY

Many nuclei spin about an axis, and the spinning generates a magnetic field known as magnetic moment. Like a normal magnet bar, this magnetic moment has a north and a south pole, which is called nuclear magnetic dipole. When the nuclei are placed in a static magnetic field, most nuclei align themselves in the same direction as the static magnetic field and represent a low energy state, while the rest, aligned in the opposite direction from the static field, are in a high energy state. The net magnetization at equilibrium is proportional to the difference in population between the two energy states. When a second magnetic field, normally in the form of radio frequency (RF) pulse, is applied to this equilibrium system, the nuclei will be excited. After excitation, the nuclei tend to return to their equilibrium states, or equilibrium population distribution. The majority of the upward transition population, originally the lower energy level population, returns to its equilibrium state by losing energy in the form of an RF wave via various radiationless transition processes termed relaxation processes. The RF wave signal is characterized by the Larmor frequency of the nuclei, and can be received and recorded by the NMR instrument (RF receiving antenna).

There are two kinds of relaxation processes: spin-lattice (longitudinal) relaxation and spin-spin (transverse) relaxation. The time constants that describe these exponential relaxation processes are known as relaxation times. The spin-lattice relaxation time is denoted by T_1 and the spin-spin relaxation time by T_2. As illustrated in Figure 1, the longitudinal relaxation process follows a recovery or growth curve and the transverse relaxation process follows a decay curve. Both relaxation processes reach their equilibrium state after at least five repetitions of their relaxation time constants.

Relaxation time is a function of the spin species and the chemical and physical environments surrounding the spins. In other words, the relaxation time constants are a fundamental property of the chemical and physical environment. The relaxation rate is related to the physical states of polymers. Therefore, analysis of T_1 and T_2 of a sample permits the study of chemical and physical properties of food polymers. Relaxation time constants should not be confused with relaxation rate. The relationship between relaxation rate R and time constant is simple and takes the form of

$$R_l = \frac{1}{T_1}, \tag{1}$$

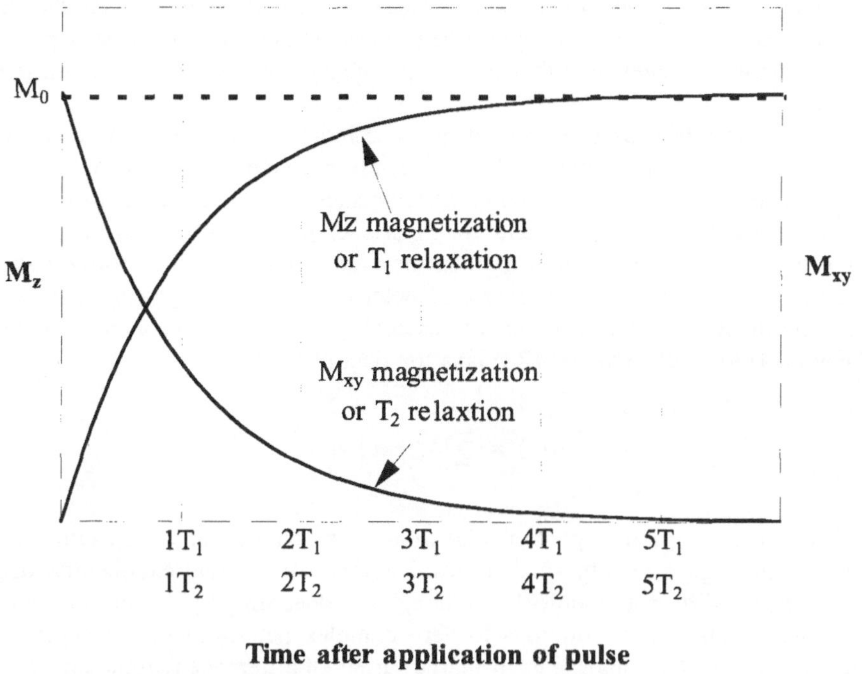

FIGURE 1 Relaxation of NMR signals after a 90° RF pulse. From Ruan and Chen.[5]

where l takes the value of 1 and 2, for spin-lattice relaxation and spin-spin relaxation, respectively.

MEASUREMENT OF T_1 AND T_2

NMR signals are produced by a series of RF pulses or magnetic field gradients (pulse sequence). A common pulse sequence for T_1 measurement is the inverse-recovery pulse sequence. The relationship between the T_1 and time can be established through the following equation

$$M_z(\tau) = M_0\left(1 - 2e^{-\frac{\tau}{T_1}}\right) \tag{2}$$

T_2 describes the decay of transverse signals, and is readily detectable. A 90° RF pulse sets the magnetization signal to its maximum value, which begins to decay immediately after the 90° pulse. The decay curve is normally termed free induction decay (FID), and can be described by the following equation

$$M_{xy} = M_0 e^{-\frac{t}{T_1}} \tag{3}$$

Other common pulse sequences for T_2 measurement are spin-echo and the Carr-Purcell-Meiboom-Gill (CPMG) pulse sequences. These pulse sequences produce more accurate T_2 values than the simple 90° pulse in an inhomogeneous magnetic field.

Many researchers used relaxation times to indicate the state of water in food polymers such as flour dough.[6-10] It is generally believed that water molecules in food polymers are somehow bound to different sites of the chemical constituents or in exchange with bound water, and are experiencing faster relaxation than in bulk water, thus having much shorter T_2 than the bulk water. Therefore, using a single relaxation time to describe complex and heterogeneous food systems may be an oversimplification. Therefore, many researchers use multiexponential models (Equation 4) to analyze the NMR relaxation data of foods.

$$A(t) = \sum A_{0i} \exp\left(-\frac{t}{T_{2i}}\right) \tag{4}$$

Most of these models predict relaxation decay data based on specific model assumptions, e.g., a certain small integral number of discrete exponential decay components of different mobility.[6,11] Although this does simplify the analysis, it may not produce satisfactory solutions to very complex heterogeneous systems. In a heterogeneous system, spins exist in a large variety of different environments, giving rise to a wide range of relaxation times. In addition, chemical and diffusive exchange, an important factor affecting spin-spin relaxation process, would also give rise to a variety of T_2 values, assuming there is a wide range of exchange rates within the heterogeneous system.[12-14] Therefore, the measured relaxation decay is a sum of contributions from all spins which may have sampled many different environments, and exchange with other spins at different rates during the course of the NMR experiment.[15] It is thus reasonable to assume that a continuum of relaxation times would arise from a continuum of different environments and exchange rates. One of the advantages of the continuum approach is that it is consistent with the continuum nature of food systems. Furthermore, additional information may be obtained from continuum models. There are often several peaks on a T_2 spectrum. A spectrum with a larger number of peaks or broader peaks would be expected for heterogeneous samples rather than for more homogeneous samples. Therefore, the number and degree of variation of the peaks could be used to indicate the homogeneity of the sample under analysis.

To analyze NMR relaxation data using a continuum model, sophisticated computer programs have to be used. One of such programs widely used is CONTIN by Provencher.[16,17] CONTIN is a Fortran program for inverting noisy linear operator equations. This program uses a general purpose constrained regularization method, which finds the simplest solution that is consistent with prior knowledge and the experimental data.

Ruan et al.[18] used this technique to analyze NMR relaxation properties of a flour dough. Figures 2 and 3 show T_2 spectra of the flour dough obtained using the CONTIN program. The *x*, *y* and *z* axes are moisture content, T_2 value, and amplitude,

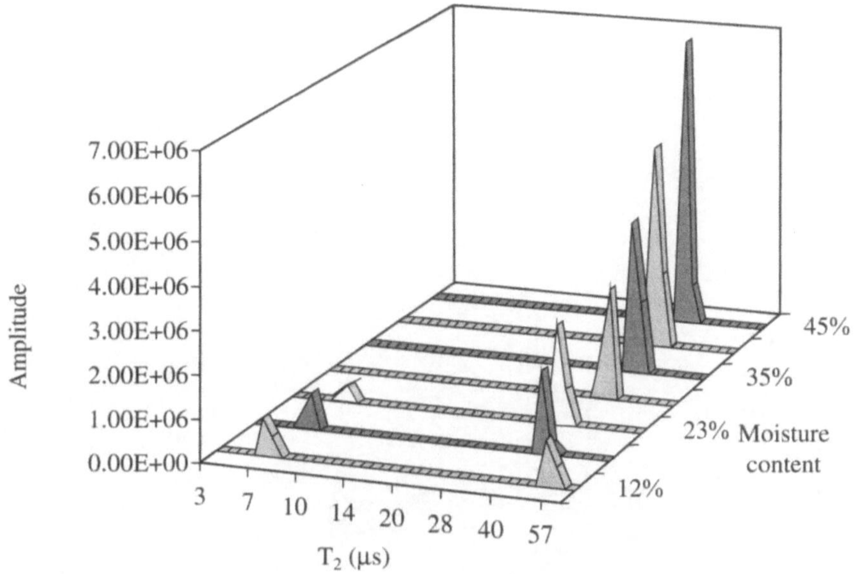

FIGURE 2 Continuous distribution of spin-spin relaxation times determined by the 90° pulse experiment as a function of moisture content. Peaks on each curve at moisture contents of 12, 18, and 23% are named as P1 and P2 from left to right, respectively, and above moisture content of 23%, the single peak on the curve is named as P2. From Ruan and Chen.[5]

respectively. Figure 2 shows the T_2 spectra computed from data obtained from the 90° pulse experiment. At moisture contents of 12 to 28%, two peaks appear on each spectrum. Water molecules covered by these two peaks can be regarded as two groups having distinctly different mobility. Because the T_2 values of these two groups range from 1 to 66 μs, proton signals falling into these two groups originated from the solids (proteins and carbohydrates) or from water molecules very close to the solids. Below 23% moisture content, the increase in the area of the second peak (T_2 > 20μs) and decrease in average T_2 values of individual peaks could be attributed to the increased available binding sites of the swollen flour substrates as a result of addition of water.[19] The disappearance of the first peak at moisture above 23% may be due to mobilization of polymers by added water. The increase in both peak area and T_2 above moisture content of 23% suggests that at the 23% moisture level all the water binding sites on the flour solids have been hydrated. Any additional water would be two or three layers away from the flour binding sites, and would therefore exchange and relax more slowly than the water molecules in the inner sites.

The CPMG pulse sequence was intended to detect proton signals having relatively longer spin-spin relaxation times than those determined by the 90° pulse experiment. The analysis of the data obtained from the CPMG experiments (Figure 3) indicated that at moisture content below 18%, no signal was detected, suggesting that the dry flour had little mobile water. At and above the 18% level, there were one to three peaks in the spectra (Figure 3). T_2 values shown in Figure 3 range from 10^2 to 10^5μs, suggesting that the detected signals were from water

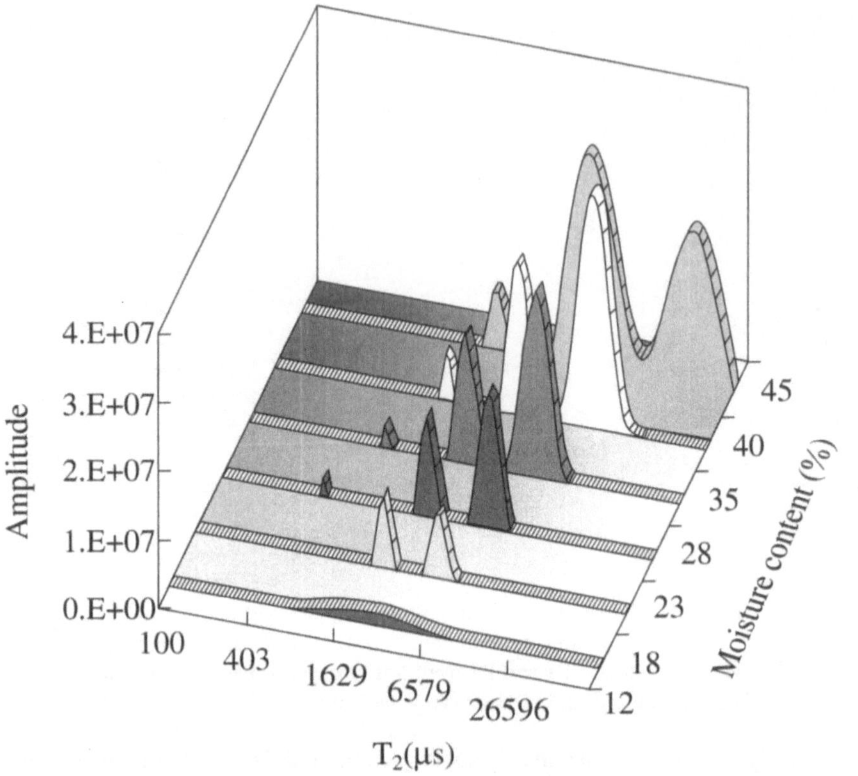

FIGURE 3 Continuous distribution of spin-spin relaxation time determined by the CPMG experiment as a function of moisture content. At and above moisture content of 23%, peaks on each curve are named as P3, P4, and P5 from left to right, respectively. From Ruan and Chen.[5]

molecules with relatively high mobility. The number and size of peaks increased, and the mean T_2 values shifted to the right (increasing T_2 value) with moisture content increasing to 28%. The appearance of new peaks suggests that new physical and chemical environments were formed within the system as a result of addition of water to the system. This coincides with the beginning of dough formation at the moisture level above 23%.

Moisture content of the dough samples affects the shape of the spectra. A broader distribution could indicate a greater variation in the chemical and physical properties of the system. Calculation of coefficient of variation (*c.v.*) of individual peaks would thus provide information about the homogeneity of the systems under analysis. The following equations were used to compute the coefficients of variation of spin-spin relaxation times[2]:

$$SD = \sqrt{\frac{\sum S_i(T_{2i} - \overline{T}_2)^2}{\sum S_i - 1}}, \tag{5}$$

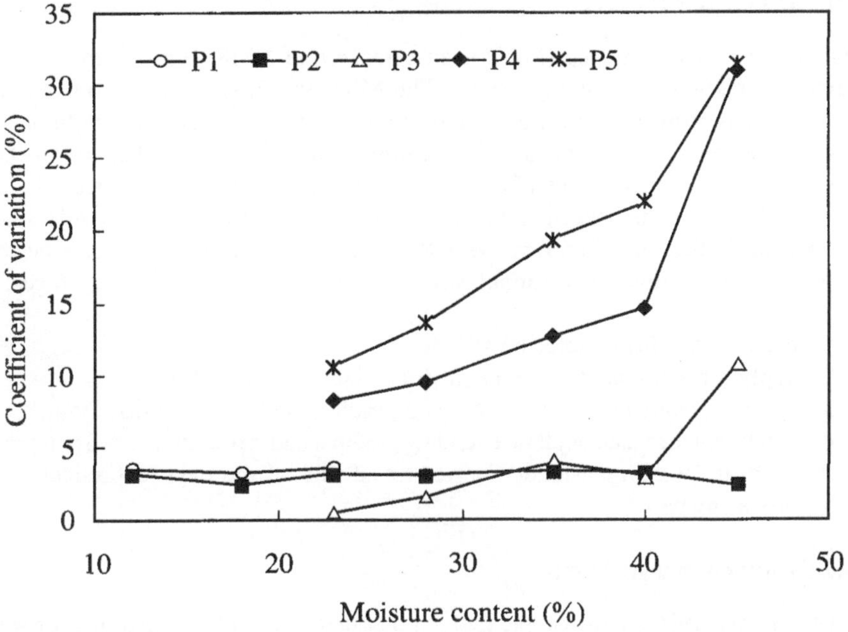

FIGURE 4 Coefficient of variation of spin-spin relaxation times determined using the continuum model (see Figures 3 and 4 for definitions of P1, P2, P3, P4, and P5).

$$c.v.\% = \frac{SD}{\overline{T_2}} \times 100 \tag{6}$$

where T_2 and S are spin-spin relaxation time and amplitude, respectively; SD is standard deviation of T_2; and $\overline{T_2} = \frac{\sum T_2 S_i}{\sum S_i}$, is the weighted average of T_2. The results are shown in Figure 4.

Figure 4 shows that the coefficient of variation of T_2 values in the range of 5 to 60 μs (the two peaks shown in Figure 2) remained almost constant as the moisture content was increased. This suggests that the environments, with which the solid-like and tightly bound water protons were associated, did not change very much, while the moisture content increased substantially.

The coefficients of variation of the longer T_2s (the three peaks shown in Figure 3) show a slow increase up to a moisture content of 40%, which may be a result of gradual formation of the bicontinuous network structure of dough and uneven distribution of water among the flour constituents, as discussed earlier. Upon reaching moisture content of 40%, which is over the normal dough moisture level, the coefficients of variation rose sharply, suggesting that the excess water may have created a very inhomogeneous dough morphology.

MRI METHODOLOGY

Magnetic resonance imaging (MRI) is an extension of NMR. It provides additional spatial information regarding the spins. The MRI system is designed to excite and receive signals from a single point, line, plane, or three-dimensional volume in a sequential manner. The point and line scanning methods are inefficient and have been superseded by the more efficient two- and three-dimensional methods. The two- and three-dimensional methods apply linear field gradients to provide spatial information. Rather than describe the MRI principles in detail, the basic procedure for production of a two-dimensional MR image is summarized into five stages:

1. excite the spins of selected slice
2. apply a phase-encoding gradient for a fixed period of time
3. apply a frequency-encoding or read gradient and collect n data points
4. increment the value of phase encoding gradient and repeat steps 1 to 3 m times
5. perform two-dimensional Fourier transform on the data to produce an $m \times n$ image

MRI STUDY OF BAKED SYSTEM

Prior to the mid 1990s, most people thought that bread staled because of loss of water. In 1852, Boussingault[21] demonstrated that hermetically sealed bread still firmed during storage, and thus opposed the moisture loss theory. However, some believe that loss of moisture from the loaf may accelerate reactions leading to bread staling.[22] Therefore, the effect of moisture redistribution, such as migration of moisture from one part of the bread to another on firming must be considered. The best technique of studying moisture movement in bread is MRI. Unfortunately, this type of study cannot be found in the literature. A relevant study by Ruan et al., in which moisture distribution in sweet roll was monitored by MRI during a five-day storage, demonstrated that moisture migrated from crumb to crust (Figure 5), which may be attributed to firming of the crumb.

NMR STUDY OF MOBILITY OF PROTONS IN BREAD

Many researchers believe that bread staling may involve starch retrogradation, interactions between starch and gluten, and glass transition. All these processes will

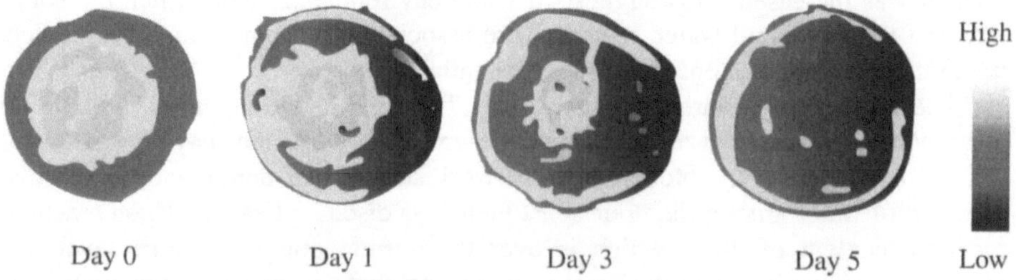

FIGURE 5 Redistribution of moisture in a sweet roll during storage at room temperature (Sample size: 5×5 cm^2). Modified from Ruan and Chen.[5]

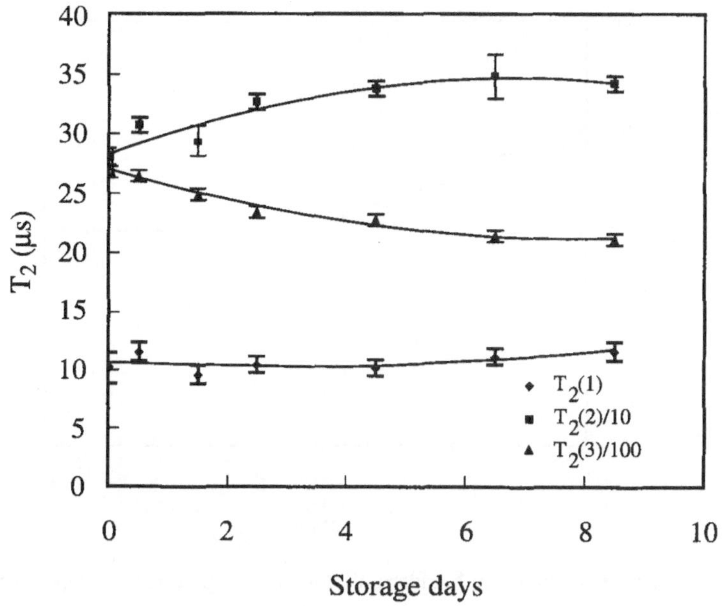

FIGURE 6 Change in T_2 values in breadcrumb during storage at room temperature.

certainly alter the chemical bonding and physical structure of bread, which will in turn affect the mobility of protons in bread.

To study the changes in the properties of protons in bread during storage, a three-exponential model was used to analyze the data collected using a low-field proton NMR[4] which classified the detected protons in the bread samples into three fractions, each of which has a distinct T_2 value range. Fraction 1 had T_2 values below 15 μs. Fraction 2 had T_2 values of 280 to 360 μs. Fraction 3 had T_2 values over 2000 μs. The T_2 values of fractions 1 and 2 exhibited a slight increase while those of Fraction 3 declined substantially over storage time (Figure 6). A decrease in overall mobility of the system over the storage time is illustrated by change in T_1 (Figure 7). The proton intensity of Fraction 1, which is related to relatively immobile, or solid, polymers, increased with increasing storage time (Figure 8). On the other hand, Fraction 3, the very mobile component, which is probably related to the liquid phase protons mainly associated with free water, decreased with increasing storage time (Figure 8). This increase in solid component and decrease in liquid component may be responsible for the increase in rigidity of bread. Fraction 2, representing the protons with intermediate mobility compared with Fractions 1 and 3, did not show apparent changes over time. However, this does not suggest that no changes occurred to this fraction of protons, since spin exchanges are possible.

MRI Study of Mobility of Protons in Baked System

The above NMR study suggests that though the mobility of the overall bread system decreased as storage time increased, some protons within the system actually experienced increased mobility. The MRI study of proton mobility of the sweet roll

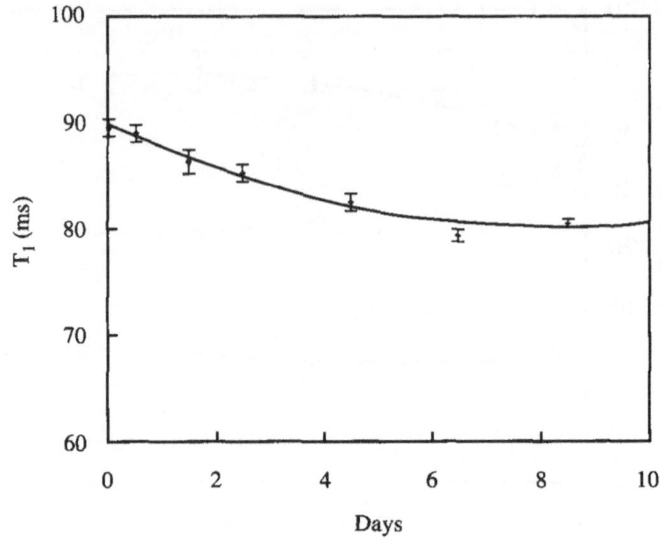

FIGURE 7 Change in T_1 values in breadcrumb during storage at room temperature.

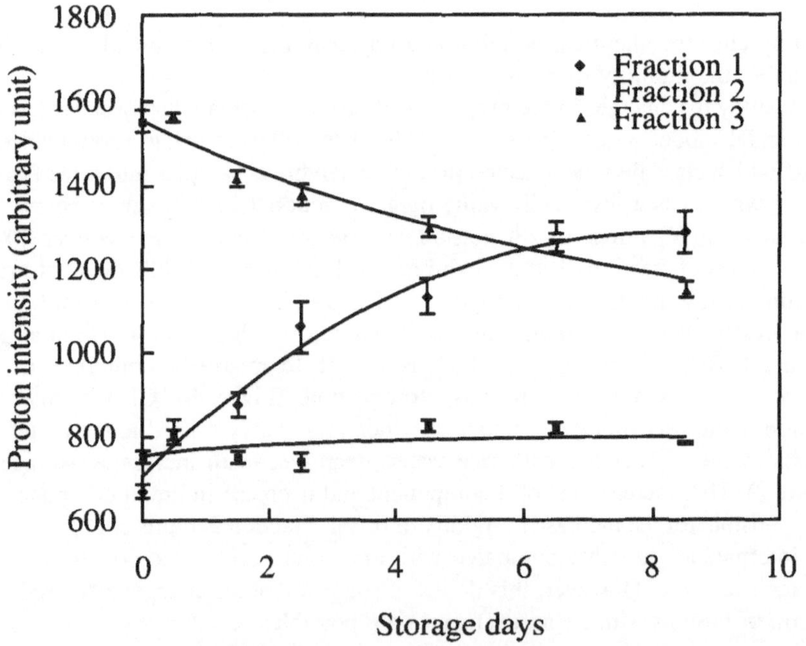

FIGURE 8 Change in proton intensity in breadcrumb during storage at room temperature.

system that was discussed earlier demonstrated that the initially uniform and low T_2 values for the central region (crumb) became higher and less uniform after the system became firmer, even though the moisture in this region was lost to the outer

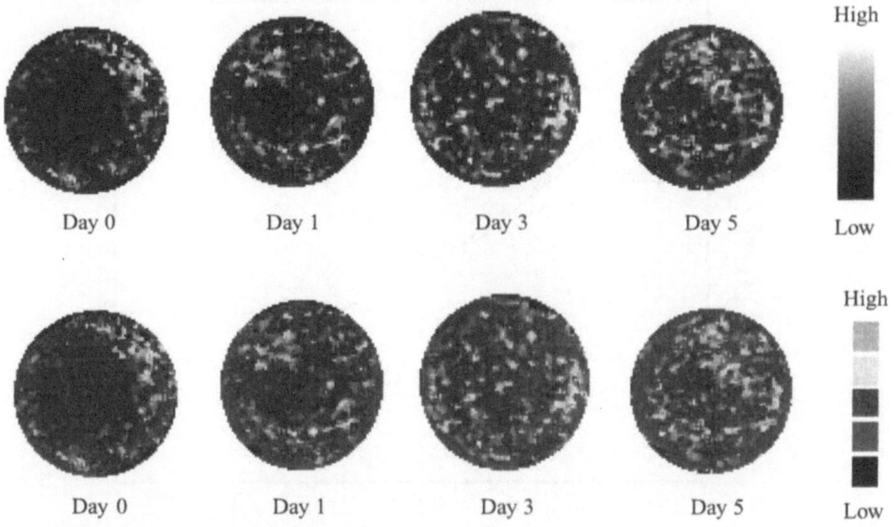

FIGURE 9 T_2 distribution in a sweet roll during a five-day storage at room temperature (Sample size: 4.2×4.2 cm^2). Modified from Ruan and Chen.[5]

region (crust) during storage (Figure 9). The increased mobility may be caused by release of water molecules from retrograding starch or other polymers such as gluten. These released water molecules disassociated themselves from the polymers and therefore had reduced plasticization, leading to a more rigid structure.

Glass Transition

The glass transition process has been demonstrated to strongly influence the physical structure and texture of food polymers. We have reported that an analysis of proton mobility as a function of temperature allows determination of glass transition temperature (T_g) of food polymers. A polymer generally consists of a number of chain segments. When temperature is increased, some segments within the long chain of the polymer molecule are first mobilized before the whole molecule starts moving. Further heating causes the entire molecule to move and become liquid. Thus we are dealing with two types of motions: internal (segmental) and external (molecular) motion.

The relationship between temperature and relaxation times is illustrated in Figure 10. T_2 changes little below a certain temperature, but after passing this temperature T_2 increases rapidly with rising temperature, suggesting a transition from the immobile state to the mobile state. For spin-lattice relaxation, there is a T_1 minimum in the temperature-T_1 curve. This peculiar behavior is related to the physical state of the system in which relaxation is taking place. In the liquid state, a long relaxation time constant indicates high mobility, and relaxation time increases with increasing temperature. When the system is moving towards solid state, for instance, due to suppressing of temperature, the spin-lattice relaxation process

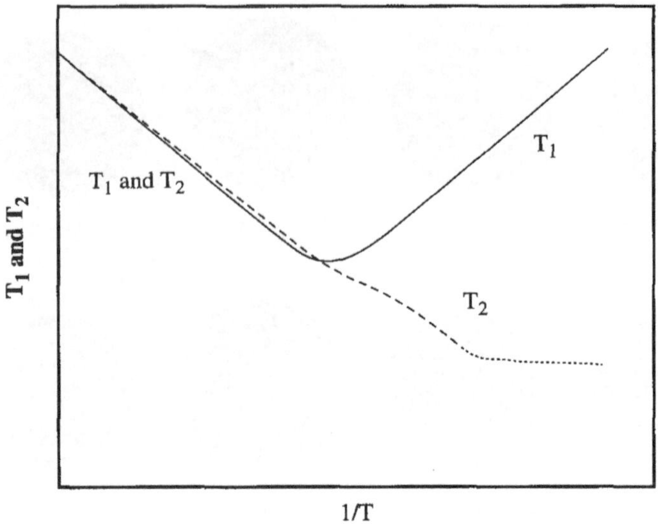

FIGURE 10 NMR relaxation time constants as a function of reciprocal temperature. From Ruan and Chen.[5]

behaves differently. The spin-lattice (T_1) relaxation is a process of energy loss by the excited spins to the lattice (the entire molecular system). To facilitate an efficient energy release by the excited spins, the lattice has to oscillate at the resonant frequency ω_0. In the solid state, the number of molecules oscillating at the resonant frequency ω_0 is very small, suggesting that a very inefficient spin-lattice relaxation process, characterized by a long T_1, can be expected. This indicates that the spin-lattice relaxation behaves markedly differently in various states, and that the temperature corresponding to the T_1 minimum may also mark a state transition temperature.

Figure 11 shows a temperature-T_2 for a bread sample. As temperature increased, T_2 did not change greatly until temperature reached around –15°C. This temperature, corresponding to the turning point on the curve, is similar to the TMA-determined glass transition temperature (–12°C) of a bread sample (37.4% moisture content) reported by LeMeste et al.[23]

Representation of a system using an average value of state transition temperature will be inadequate if the system is not sufficiently homogeneous in terms of physical structure and chemical composition. In fact, many food systems are highly heterogeneous, and often have multiple phases or components. For baked goods, a nonuniform distribution of T_g values could mean that the textural properties and stability of the product are also not uniform. For other products, in which chemical and microbiological activities are the key deteriorating factors that are strongly dependent on T_g, a nonuniform distribution of T_g values could be a major challenge to a product's safety. A localized low T_g at certain locations within a system can cause a condition well above the average T_g of the system under normal storage conditions,

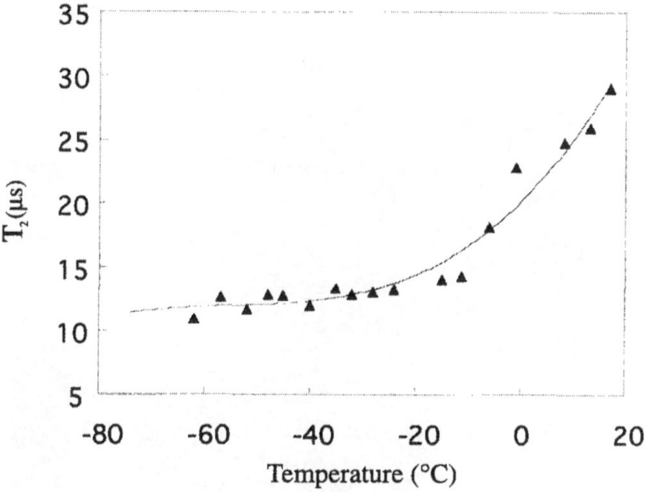

FIGURE 11 T_2 of bread as a function of temperature.

allowing chemical and microbiological activities to take place. Therefore, an evaluation of the distribution of the T_g values within a heterogeneous system would allow more reliable prediction of physical properties, and stability of the entire system.

Because MRI can measure all the parameters that an NMR spectroscopy can, MRI can also produce T_1 minimum ($T_{T_{1min}}$) or T_2 turning point (T_{T_2}) to indicate the glass transition temperatures. In addition, MRI can provide information regarding the spatial distribution of these parameters, and thus allow construction of images or maps to illustrate distribution of T_1 and T_2. If a series of maps of T_1 or T_2 distribution in a food system at different temperatures is obtained, we will be able to determine the $T_{T_{1min}}$ or T_{T_2} of any points in the system (any pixels in the image). With these data, a map of $T_{T_{1min}}$ or T_{T_2} can be constructed.

T_1 maps are relatively easy to acquire with current MRI instrumentation. A three-step method can be followed to produce $T_{T_{1min}}$ maps.[24] In the first step, a series of T_1 images of the same sample at different temperatures is obtained using MRI. Each pixel of the image has a T_1 value at every temperature tested. In the second step, $T_{T_{1min}}$ for each pixel is computed using a fitting program. In the final step, an image consisting of $T_{T_{1min}}$ for all individual pixels is reconstructed and is regarded as a state transition temperature map.

Figure 12 is a T_g map for a bread sample. The average value is −10°C, with a standard deviation 11°C. This result is similar to the NMR-determined value (−15°C).

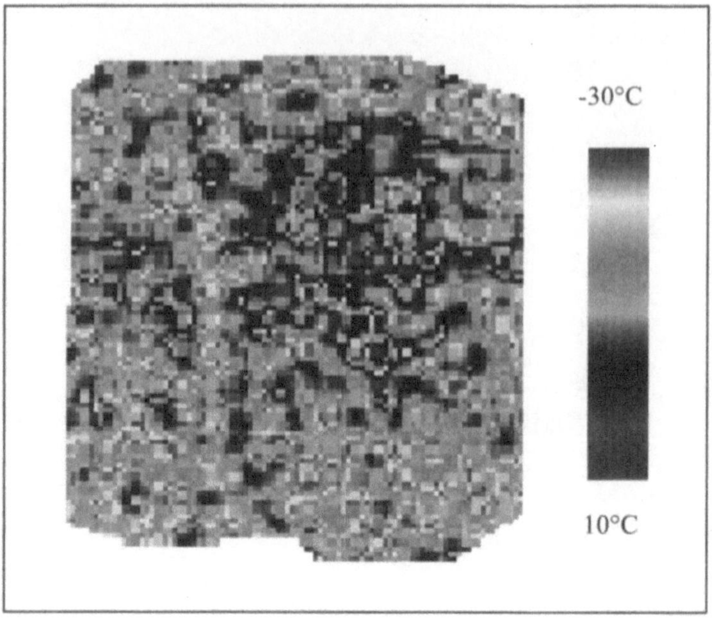

FIGURE 12 T_g map of a bread sample (Sample size: 2.5 × 2.5 cm^2, moisture content: 39.5% g/g, wet basis).

REFERENCES

1. Seow, C. C. and Teo, C. H., Staling of starch-based products: a comparative study by firmness and pulsed NMR measurements, *Starch,* 48, 90, 1996.
2. Kim-Shin, M. S., Mari, F., Rao, P. A., Stengle, T. R., and Chinachoti, P., 0-17 nuclear magnetic resonance studies of water mobility during bread staling, *J. Agric. Food Chem.,* 39, 1915, 1991.
3. Morgan, K. R., Gerrard, J., Every, D., Ross, M., and Gilpin, M., Staling in starch breads: the effect of antistaling alpha-amylase, *Starch,* 49, 54, 1997.
4. Chen, P. L., Long, Z., Ruan, R., and Labuza, T. P., Nuclear magnetic resonance studies of water mobility in bread during storage, *Lebensm. Wiss. Technol.,* 30, 178, 1997.
5. Ruan, R. and Chen, P. L., *Water in Food and Biological Materials: A Nuclear Magnetic Resonance Approach,* Technomic Publishing, Lancaster, 1998.
6. Leung, H. K., Magnuson, J. A., and Bruinsma, B. L., Pulsed nuclear magnetic resonance study of water mobility in flour doughs, *J. Food Sci.,* 44, 1408, 1979.
7. Belton, P. S., Colquhoun, I. J., Grant, A., Wellner, N., Field, J. M., Shewry, P. R., and Tatham, A. S., FTIR and NMR studies on the hydration of a high-Mr subunit of glutenin, *Int. J. Biol. Macromol.,* 17, 74, 1995.
8. Leung, H. K., Magnuson, J. A., and Bruinsma, B. L., Water binding of wheat flour doughs and breads as studied by deuteron relaxation, *J. Food Sci.,* 48, 95, 1983.
9. Richardson, S. J., Baianu, I. C., and Steinberg, M. P., Mobility of water in wheat flour suspensions as studied by proton and oxygen-17 nuclear magnetic resonance, *J. Agric. Food Chem.,* 34, 17, 1986.

10. Toledo, R., Steinberg, M. P., and Nelson, A. I., Quantitative determination of bound water by NMR, *J. Food Sci.*, 33, 315, 1968.
11. Schmidt, S. J. and Lai, H., Use of NMR and MRI to study water relations in foods, in *Water Relationships in Foods,* Lavine, H. and Slade, L., Eds., Plenum Press, New York, 1990, 405.
12. Zimmerman, J. R. and Brittin, W. E., Nuclear magnetic resonance studies in multiple phase systems: lifetime of a water molecule in an adsorbing phase on silica gel, *J. Phys. Chem.*, 61, 1328, 1957.
13. Lillford, P. J., Clark, A. H., and Jones, D. V., Distribution of water in heterogeneous foods and model systems, in *ACS Symposium Series, Water in Polymers,* Rowland, S. P., Ed., American Chemical Society, Washington, 1980, 177.
14. Belton, P. S. and Hills, B. P., The effect of diffusive exchange in heterogeneous systems on NMR line shapes and relaxation processes, *Molec. Phys.*, 61, 999, 1987.
15. Kroeker, R. M. and Hendelman, R. M., Analysis of biological NMR relaxation data with continuous distribution of relaxation times, *J. Magn. Reson.*, 69, 218, 1986.
16. Provencher, S. W., A constrained regulation method for inverting data represented by linear algebraic or integral equations, *Comput. Phys. Commun.*, 27, 213, 1982.
17. Provencher, S. W., CONTIN: A general purpose constrained regularization program for inverting noisy linear algebraic and integral equations, *Comput. Phys. Commun.*, 27, 229, 1982.
18. Ruan, R. R., Wang, X., Chen, P. L., and Wang, X., NMR study of water in dough, ASAE Paper 976062, in *ASAE Annual International Meeting,* Minneapolis, Minnesota, USA, 1997.
19. Bushuk, W. and Mehrotra, V. K., Studies of water binding by differential thermal analysis. II. Dough studies using the melting mode, *Cereal Chem.*, 54, 320, 1977.
20. Devore, J. L., *Probability and Statistics for Engineering,* Brooks/Cole Publishing, Monterey, CA, 1982.
21. Boussingault, J. B., Experiments to determine the transformation of fresh bread into staled bread, *Ann. Chem. Phys.*, 36, 490, 1852.
22. MacMasters, M. M., Starch research and baking, *Bakery Dig.*, 35, 42, 1961.
23. LeMeste, M., Huang, V. T., Panama, J., Anderson, G., and Lentz, R., Glass transition of bread, *Cereal Foods World,* 37, 264, 1992.
24. Ruan, R. R., Long, Z., Chang, K., Chen, P. L., and Taub, I. A., Glass transition temperature mapping using magnetic resonance imaging, *Trans. ASAE,* 42, 1055, 1999.

7 Thermo-Mechanical Behavior of Concentrated Starch-Based Products

Martine Le Meste, Eleni Chiotelli, and Arnaud Rolée

CONTENTS

INTRODUCTION

Technological functionality of the main ingredients of a flour depends on the processing stage considered. In bread or cookie dough, starch is often considered a relatively inert filler dispersed in the viscoelastic gluten network. This network

0-8493-8790-6/01/$0.00+$.50

controls the rheological behavior of the dough itself, such as its elasticity and extensibility during fermentation, machining (sheeting and laminating), relaxation, cutting, and thermal expansion during baking. Starch granules exhibit swelling during the baking process. Although swelling is incomplete in conditions of limited hydration, this event may be the dominant parameter controlling the temperature-induced rheological changes of concentrated suspensions. During processing, the transformations of both the gluten network and starch granules strongly depend on the amount and properties of water present in the dough. Indeed, through its solvent and plasticizing properties, water controls the dough rheological properties, its behavior during cooking, texture of the finished product, and its evolution during storage. The rate of bread staling may depend on the transformations occurring during baking and on the overall structure of the preparation at the end of the cooking process.

The objective of this chapter is to describe physical transformations occurring during baking, with special emphasis on the behavior of starch and the role of water. However, we also studied the influence of other ingredients present in most dough (sucrose, oil, gluten). We used differential scanning calorimetry (DSC) to measure loss in starch crystallinity during heating. The consequences of the temperature-induced structural changes (granule swelling, melting of starch crystallites) on the rheology of starch suspensions were studied using the dynamic thermo–mechanical analysis (DMTA). Changes in the properties of the aqueous phase induced by the increase in temperature, and the subsequent structural changes were investigated using the spin probe electron spin resonance (ESR) method.

METHODS

A brief discussion of the experimental procedures used in our studies is presented here. More detailed information is available elsewhere in this book.

Sample Preparation

Wheat and waxy maize starches were provided by Roquette Frères (Lestrem, France). Preparations were made with 18g of starch (dry matter: 15.75g of wheat starch or 16.06g of waxy maize starch). Distilled water was added until moisture contents of 25, 30, 35, 40, 45, 50, 55 and 60% (w/w wb) were obtained. Manual blending was continued until a homogeneous mixture was obtained. The mixture was then let to rest for at least one hour at room temperature in a closed environment. The more liquid-like samples were magnetically stirred.

DSC

Thermograms were obtained with a Perkin-Elmer DSC-7 differential scanning calorimeter, equipped with a TAC/7 DX thermal analysis data station (Perkin Elmer, France), calibrated with azobenzene and indium in the positive temperature range. Fractions of starch preparations (40 to 85 mg) were weighed and hermetically sealed in stainless steel DSC pans. All traces were normalized to 1 mg of starch. The

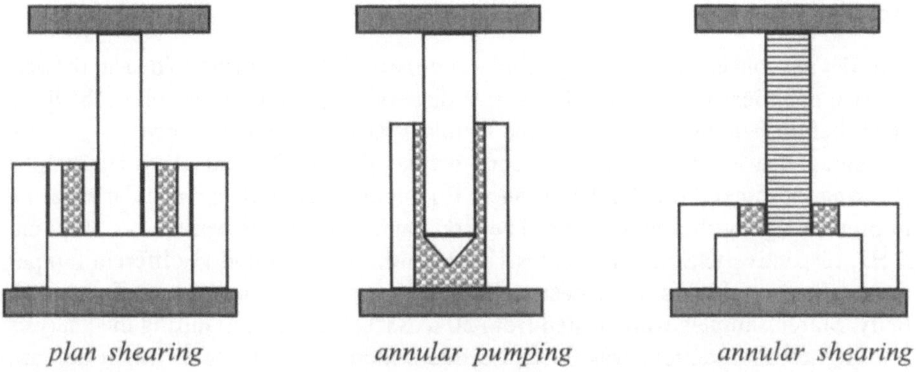

FIGURE 1 Schematic cross sections of the different devices used for DMTA.

scanning temperature range and the heating rate were 25 to 160°C and 10°C/min, respectively. An empty pan was used as an inert reference. All tests were performed at least in triplicate. The partial melting enthalpy was calculated from the onset of the endotherm to 85°C (per 1°C step) to plot the curve representing the cumulated enthalpy values versus temperature. The average standard deviation was calculated to be 8.9%.

DMTA

The small amplitude oscillatory rheological measurement was performed with a viscoanalyzer (Metravib R.D.S, France), equipped with a thermocontrol unit. The temperature was monitored by a thermoprobe at ±0.5°C. Plan shearing was used for the more solid-like samples. For the more liquid-like samples, two different devices were used, depending on the range of temperature: annular pumping up to approximately 63°C, then annular shearing up to 85°C. This change in device was necessary because during the thermal treatment, liquid-like samples became more rigid and annular pumping was no longer efficient. These different modes are showed in Figure 1.

Plan Shearing

Two samples of the same size were used for this device. Samples were 3 mm high × 15 mm diameter or 4 mm high × 20 mm diameter. They were vertically glued with cyanoacrylate glue (Amatron, England) on outside and inside plates. The inside plates were connected to a sensor which regulated amplitude and frequency of the strain, and the outside plates were connected to a sensor which registered stress.

Annular Pumping

Sample was poured into a cylindrical cell. A piston oscillated with small amplitude in the center of the cell, which was glued with cyanoacrylate glue onto the sensor registering stress. The piston was screwed into the sensor, which regulated amplitude and frequency of the strain.

Annular Shearing

This device consisted of coaxial cylinders connected to the sensors. The gap between the two cylinders was 2 mm. The sample dispersion, poured in the cell in the liquid state, before heating, was held in the annular space by capillary force.

Samples were coated with silicone grease (Rhône Poulenc, France) for plan shearing, or mineral oil (Nachet, France) for annular pumping and annular shearing, to prevent drying during analysis. The strain and frequencies were set at 3 μm and 5 Hz, respectively. Strain sweep tests were initially performed at different temperatures. They confirmed that measurements were run in the linear range of viscoelasticity. Starch samples were heated from 30 to 85°C (1.5°C/min) during the analysis. The highest temperature was 85°C, beyond which starch granules might be damaged.[1] The VA2000 software package provided by Metravib R.D.S allowed calculation of rheological parameters, including storage modulus (G′). All tests were performed at least in triplicate. The average standard deviation for all tests was calculated to be 10.4% for wheat starch and 10.2% for waxy maize starch.

ESR

Hydrated starch has no paramagnetic activity, so the spin-probing technique was employed, in which a compound with a nitroxide radical, possessing a stable free electron, is added to the system. The ESR spectra reflect the motion of the small paramagnetic probe that depends on probe size, solvent viscosity, and temperature. The size and polarity of the probes influence their accessibility to microenvironments and their behavior in a network. A smaller probe may stay relatively mobile for steric reasons, while a larger molecule may have reduced mobility. The 4-hydroxy,2,2,6,6-tetramethyl-piperidine N-oxyl (TEMPOL) radical was purchased from Aldrich Chemicals (Strasbourg, France). Because of its small size, this probe is able to diffuse into the starch granules, so it can be dispersed in the aqueous phase inside and outside the granules. When preparing the samples, 300μl of a TEMPOL aqueous solution (2 mg/ml) was added to distilled water, then manually mixed with starch (approximate final TEMPOL concentration: 2.44×10^{-7} mol/g of dry starch). Sealed capillary tubes containing aliquots of the samples were placed in a 3mm diameter ESR sample quartz tube, and ESR spectra were collected using a Bruker EMX spectrometer (Bruker, France) with a nitrogen-flow temperature control. The operating frequency and center field were at about 9.42 GHz and 3357 G, respectively. The spectra were recorded at a microwave power of 10 mW. Any saturation phenomenon was avoided. The scan rate, time constant, and modulation amplitude were adjusted so that distortion of the spectra was avoided. For all experiments, the temperature was varied every 2°C, between 25 and 85°C, and the sample was stabilized for three minutes before recording the spectra. All ESR experiments were carried out in triplicate. The average standard deviation for all the tests was calculated to be 8.8% for wheat starch and 8.1% for waxy maize starch.

The rotational correlation time (τ_c) was determined from the relation:

$$\tau_c = 6.65 \times 10^{-10}(\Delta H_{I+1}) \times [(I_{+1}/I_{-1})^{1/2} - 1] \qquad (1)$$

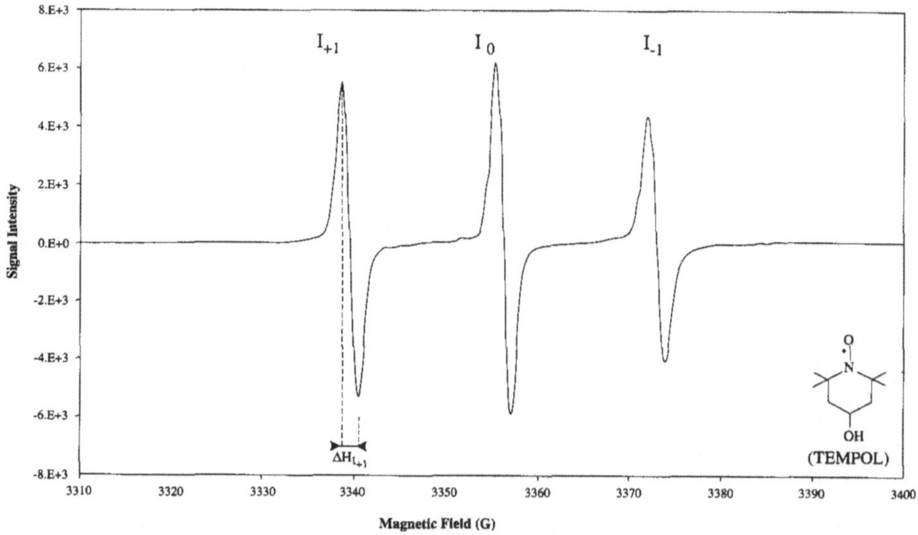

FIGURE 2 Typical ESR spectrum of a nitroxide free radical (TEMPOL). I_{+1} and I_{-1} are, respectively, the height of the lines I_{+1} and I_{-1}, and ΔH_{I+1} is the width of the I_{+1} line.

deduced from the Freed and Fraenkel[2] theory where ΔH_{I+1} is the width of the I_{+1} line and I_{+1} and I_{-1} are the height of the lines I_{+1} and I_{-1}, respectively (Figure 2).

The conventional ESR method was used, allowing mobility measurements in the range $10^{-11} < \tau_c < 10^{-7}$s. The rotational diffusion coefficient (D_{rot}) was evaluated from the rotational correlation time (τ_c):

$$D_{rot} = 1/(6\tau_c) \tag{2}$$

RESULTS AND DISCUSSION

STARCH BASED SUSPENSIONS AT ROOM TEMPERATURE

Rheological Behavior

Theoretical aspects

The viscosity, or the Coulomb (G) or Young (E) modulus, of a suspension of non-interacting rigid particles, is a function of the volume fraction occupied by dispersed particles (ϕ) and of viscosity of the continuous phase as described by the following relation:

$$\eta = \eta_s f(\phi) \tag{3}$$

with η the apparent viscosity of the suspension at a given shear rate and η_s the apparent viscosity of the continuous phase;

$$f(\phi) = 1 + 2.5\phi \tag{4}$$

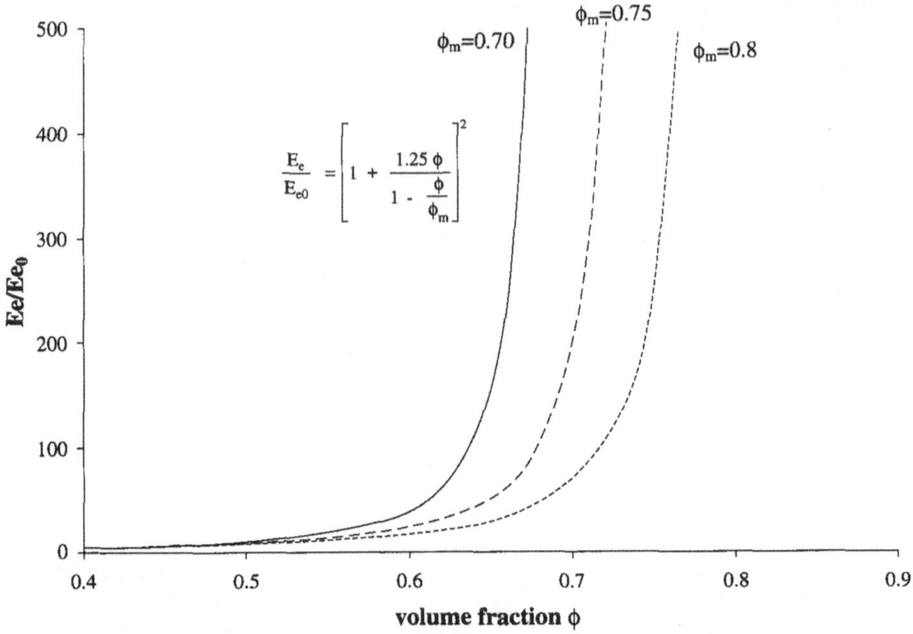

FIGURE 3 Influence of the volume fraction of particles on the relative modulus of dispersions for different values of the volume fraction at close packing ϕ_m as predicted with the Eilers and Van Dijk equation.

for dilute suspensions (Einstein Equation). For more concentrated suspensions, several relations have been proposed. Among them, the Krieger-Dougherty relation is often cited[3]:

$$\eta = \eta_s[1 - (\phi/\phi_m)]^{-2.5\phi m} \tag{5}$$

were ϕ_m is the volume fraction corresponding to the close packing of particles. Similar relations have been proposed for the G or E modulus of suspensions of rigid particles, such as the Eilers and Van Dijk equation[4,5]:

$$(E/E_0) = [1 + (1.25\phi)/(1 - \phi/\phi_m)]^2 \tag{6}$$

where E_0 is the E modulus of the continuous phase. This relation and its graphic representation (Figure 3) show that the main parameter controlling the modulus of a concentrated suspension of rigid particle is the volume fraction of the dispersed phase and that this modulus increases drastically when the volume fraction gets close to ϕ_m.

Starch water preparations

Starch granules from different origins, thus with different shapes and sizes, exhibit different ϕ_m values. Indeed, ϕ_m equals almost 1 and 0.74 for the regular arrangement

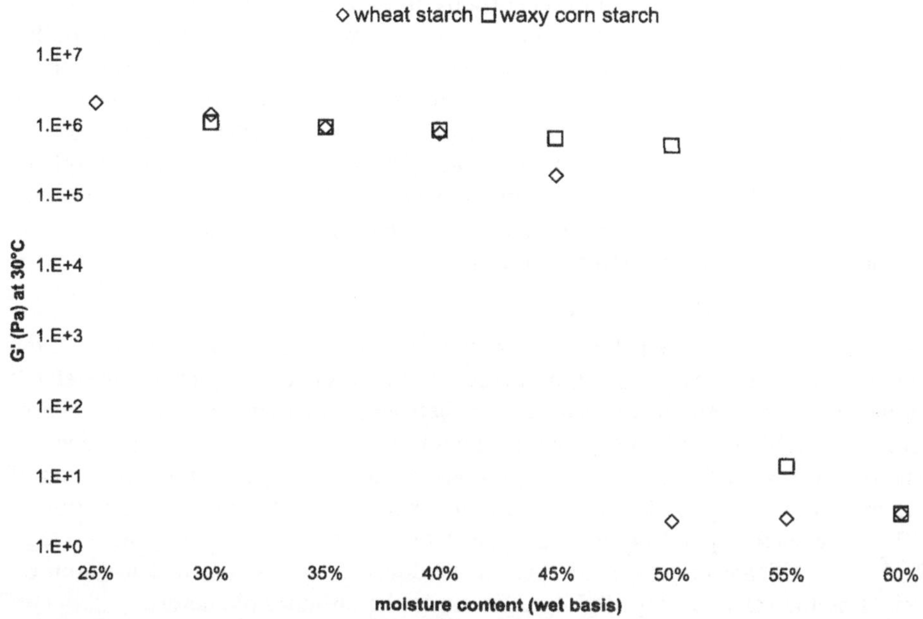

FIGURE 4 Influence of moisture content of the starch suspensions on the modulus of wheat and waxy corn starch suspensions at 30°C.

of square particles and spheres respectively, $\phi_m = 0.64$ for a random distribution of spheres, or even 0.33 for a random arrangement of irregular shape particles. ϕ_m also depends on the distribution of the granule size. For a mono-modal distribution of particle size:

$$\phi_m = 1 - 0.47\left(\frac{d}{D}\right)^{0.2} \tag{7}$$

where d and D are the lower and the upper bounds of size distribution of the particles, respectively. A previous study using laser-light diffraction[6] showed that wheat starch had a far larger size distribution than waxy corn starch, for example. Using the previous relation and the published values of granule size distribution, ϕ_m could be estimated to be 0.73 for wheat starch and 0.66 for waxy corn starch. Moreover, the ϕ_m value calculated for wheat should be even higher because of the polydispersity of the wheat granules.

It can thus be expected that the shear modulus of starch suspension might remain quite low and constant at low starch volume fraction (or mass ratio), and as ϕ_m is approached the modulus should increase drastically, as shown in Figure 3. This correlates with results shown in Figure 4 for wheat and waxy corn starches, where ϕ_m is reached at a lower starch mass fraction for waxy corn starch than for wheat starch. This could be attributed to differences in the granule sizes and shapes, and to the higher affinity for water, and thus swelling power, of waxy corn starch relative to wheat starch.

Up to ϕ_m, the modulus is not expected to be sensitive to differences in granule rigidity that may originate from differences in the structure, organization, and physical state between granules from different origins. Indeed, at high moisture contents (i.e., low granule volume fractions), the elastic modulus of both types of starch is similar. Above ϕ_m, a sensitivity of the modulus to granule rigidity is expected, however, no significant difference between the two starches is observed. In agreement with Zeleznak and Hoseney,[7] our DMTA results suggest that whatever the moisture content in the range considered in our studies (25 to 60% wet basis), the amorphous regions are in the rubbery state at room temperature. However, native granules can be considered as relatively rigid particles, even in presence of excess water, because of the presence of the crystalline regions.

Influence of the other ingredients

Adding sucrose to a given starch-water mixture increases the volume fraction of the continuous liquid phase, and thus decreases the volume occupied by the starch granules. A competition for water occurs between starch and sucrose and, consequently, swelling of the native starch granule may be altered by the presence of sucrose. The viscosity of the continuous phase may also depend on the partition of water between the granules and the continuous intergranular phase. The addition of 10 or 20g sucrose per 100g of starch-water mixture with 60% water (i.e., $\phi < \phi_m$), had no significant effect on the mixture modulus. It was assumed that starch ϕ_m might be the same or only slightly different in the presence of sucrose.

In presence of 45% water ($\phi > \phi_m$), the elastic shear modulus (G′) of wheat starch was on the order of $3\ 10^5$ Pa. At 42% moisture, both wheat dough and gluten exhibited the same significantly lower value of modulus (on the order of $1\ 10^4$ Pa). This effect may be partly attributed to the lower volume fraction occupied by the dispersed phase (starch granules) ($\phi < \phi_m$), and to the fact that when ϕ becomes lower than ϕ_m, the rheology of the dispersion may be controlled by the viscoelasticity of the continuous polymeric phase.

Triolein (10 or 18% w/w wet starch) added to a 48% potato starch suspension (ϕ close to ϕ_m) also induced a decrease in G′, again associated with decrease in starch volume fraction.

We can thus conclude that the rheological behavior of a native starch-based preparation, at room temperature, is very sensitive to small changes in the volume fraction of the dispersed granules when this volume fraction ϕ is close to its ϕ_m value. ϕ depends on the starch concentration and the affinity of the native granules for water and, thus, on their swelling ability at room temperature. ϕ_m is controlled by the shape, size, and size distribution, i.e., by the botanical origin of starch. When $\phi < \phi_m$, the continuous aqueous phase controls the rheology of the suspension. When $\phi > \phi_m$, the granule rigidity or deformability might play a determinant role. This will be discussed later in this chapter.

Mobility of Water and Water Soluble Solutes

At room temperature, even in the presence of excess water, starch hydration and the subsequent swelling of granules remains relatively limited (around 5% for wheat starch).[8] NMR studies have shown that water molecules in starch are highly mobile, even at low water contents,[9] and can diffuse rather rapidly into the starch samples.

Water molecules have the possibility to penetrate inside the granules and then interact with the starch chains located within the amorphous domains, however, the rate of swelling appears to be controlled by the presence of the crystalline lamellar domains. Water mobility may vary, depending on its partitioning among domains. Moreover, changing the initial water content and temperature may influence this distribution of the water within the different phases or domains.[10]

ESR proved to be an appropriate technique to detect changes in the properties of the water phase within starch dispersions.[11,12] It measures rotational mobility of water-soluble spin probes (TEMPOL) dispersed in the aqueous medium of the suspension. The rotational diffusivity of a probe dispersed in a homogeneous liquid medium can be described by the modified Stokes-Einstein equation:

$$D_{rot} = \frac{kT}{8\pi\eta r^3 C} \qquad (8)$$

where k is the Boltzmann's constant, T the absolute temperature, η the viscosity of the medium, r the radius of the diffusing molecule, and C the coupling parameter representing the amount of solvent that is dragged with the molecule when it moves.[13] This factor is generally close to 1 in aqueous solutions, except for low moisture systems. Rolée and Le Meste[12] emphasized that, in heterogeneous systems such as concentrated starch-water dispersions, the relevant parameter controlling the mobility of water soluble solutes is the viscosity of the diffusion medium (i.e., of the aqueous phase). We hypothesized that the molecular mobility of the water-soluble probe should be controlled by the mobility of the water molecules, which depends on the interactions between water and starch molecules. This method should thus be able to detect the consequences on probe mobility of starch structural modifications, such as disruption of low energy starch-starch interactions (melting), subsequent increase in starch-water hydrogen bonding, and starch flexibility.

At room temperature, as in all the temperature ranges studied, three line spectra similar to those of probes in water were obtained with wheat starch dispersions (30 to 60% water). This behavior is generally interpreted in terms of isotropic fast motion. The calculated rotational diffusion coefficients (Figure 5) were found to be lower than the values for the probe in water. The observed slower motion in the presence of starch suggests that the probe experienced an environment of higher viscosity than in water. For both waxy corn and wheat starches, rotational diffusivity of probes increased with water content as a consequence of the plasticizing effect of water on hydrophilic solutes. Only one population of probes was observed, i.e., probes inside and outside the granules could not be distinguished. No specific probe–starch interaction (which would have been characterized by very slow probe motions) was observed.

In comparison to wheat starch, and for similar concentrations (above 40% moisture content), waxy corn starch dispersions showed lower D_{rot} values at room temperature, suggesting favored starch-water interactions and/or more rigid polymeric chains for waxy corn starch (Figure 5). To support these interpretations, wheat and waxy corn starch dispersions of equal weight, with excess water (90% wb), were prepared and were allowed to stand overnight in a closed environment under magnetic stirring. Both preparations were then allowed to sedimentate for 24 hours.

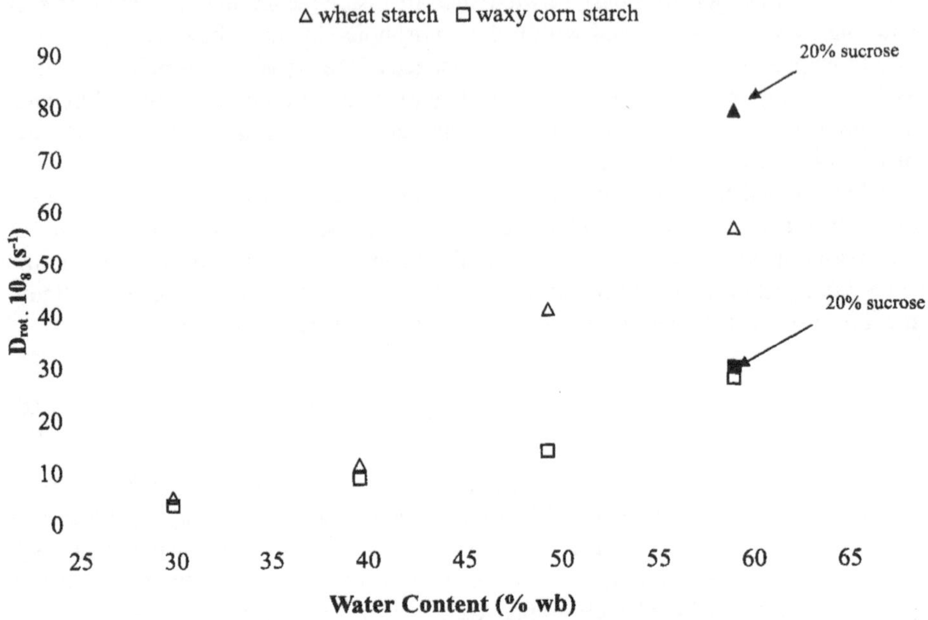

FIGURE 5 Influence of the water content on the rotational diffusion coefficient of TEMPOL dispersed in native wheat and waxy corn starch preparations with (filled symbols) and without (open symbols) sucrose added, at 25°C. The water content is expressed as the amount of water for 100 g of starch-water mixture. In presence of sucrose, the water content expressed in g water per 100 g wet total mixture is 50%.

Comparison showed clearly that the height of waxy corn sediment was greater than that of wheat starch. Thus, the affinity of waxy corn starch for water would be higher, confirming the previous ESR results.

Influence of the other ingredients

Different amounts of sucrose (10 or 20g sucrose for 100g of starch-water preparations) were added to 60% (w/w wet basis) starch–water preparations prepared with wheat or waxy corn starch. Probe mobility in both mixtures was lower in the presence of starch than in sucrose solutions that had the same sucrose concentration, but without starch. Starch again is shown to increase the local viscosity experienced by the probe. Whereas addition of sucrose only slightly modified the rotational mobility of the spin probe when dissolved in the water phase of the waxy corn starch preparation, it resulted in a significantly higher probe mobility within the wheat starch preparation, despite the overall reduction in water content (Figure 5). The change in viscosity of the probe environment upon sucrose addition depends on the balance between starch and water, sucrose and water, and possibly on sucrose-starch interactions and the consequences of these interactions on the mobility of each component. Both water and sucrose act as plasticizers for starch, however the mobility of the starch molecules controls the mobility of at least some of the water molecules. Changing a liquid phase from water to a sucrose solution (in absence of starch) induces an increase in viscosity and thus reduces probe mobility. The opposite was observed with starch-based preparations. This could be due to the increase in starch flexibility in the presence of sucrose and/or the influence of sucrose on starch-water interactions, where less water would be available for starch hydration in presence of 10 to 20% sucrose.

When a wheat starch suspension and a wheat dough with the same water content (42% w/w wet basis) are compared, a slightly higher probe mobility is observed in the dough: 14 10^8 s^{-1} and 18 10^8 s^{-1} for starch and dough, respectively. Adding 10 or 20% sucrose (g/100 g preparation) to the dough also slightly increased diffusivity of the water-soluble probe at room temperature.

In conclusion, this ESR study of probe mobility in native starch-based preparations at room temperature does not show any evidence of heterogeneous distribution of the water-soluble probe within the preparation. Rotational mobility, which is controlled by the local viscosity of the aqueous phase surrounding the probe, increases with moisture content for both starches, but this increase is more pronounced for wheat starch than for waxy corn starch. Thus, ESR results suggest that waxy corn starch exhibits a more pronounced effect on the dynamical properties of the water liquid phase than wheat starch. ESR results also suggest that addition of sucrose affects the flexibility of starch and/or the starch-water interactions.

Physical Transformations of Starch-Based Preparations during Heating

The physical transformations that occur during heating starch in the presence of water are complex and strongly water dependent. The gelatinization process of starch involves modifications of the starch physical state, i.e., progressive disappearance of the crystalline regions, modification of starch–water interactions, morphological changes of starch granules during the swelling process, and progressive dissolution of amylose and amylopectin, with their subsequent diffusion out of the granules. The evolution of the rheological properties of starch-based preparations during heating is the result of all these factors. If physical changes are limited in the temperature range considered because of a low water content, changes in rheological properties will also be limited. As water content increases, the extent of the physical transformations affecting the granules will increase, and the modulus of the preparation will also exhibit important changes.

Shear Modulus and Crystal Melting

The changes in the storage modulus (G′) of wheat and waxy corn starches induced by heating in presence of different water contents are shown in Figures 6 and 7. Two classes of samples can be distinguished: the first one with an initial $\phi < \phi_m$, the second one with $\phi \geq \phi_m$. For the latter class, only limited changes occur during heating. A slight increase in G′ is observed above around 50°C for wheat starch, followed by a gentle decrease above 70°C. The second class, the most hydrated samples, exhibits a much stronger increase in rigidity between 50 and 60°C for wheat starch. The modulus reaches approximately the same values as those of more

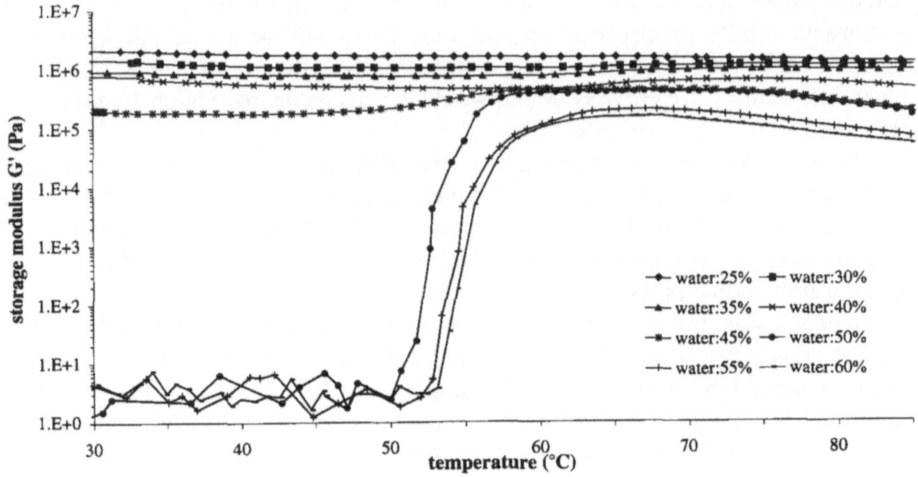

FIGURE 6 Storage modulus changes of wheat starch dispersions at intermediate moisture contents (w/w, wb), as a function of temperature during heating.

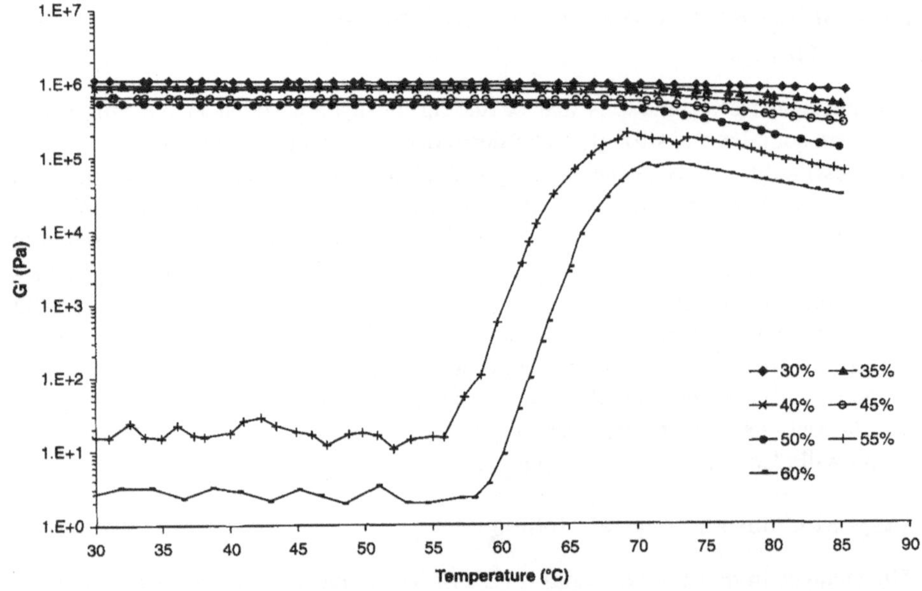

FIGURE 7 Storage modulus changes of waxy corn starch dispersions at intermediate moisture contents (w/w, wb), as a function of temperature during heating.

concentrated preparations. For waxy corn starch, only a slight decrease is observed above 65 to 70°C for the first class of less hydrated preparations. For the second class, the strong increase in rigidity was shifted to higher temperatures (around 70°C) compared to wheat starch. This increase in G′ was attributed to the granule swelling

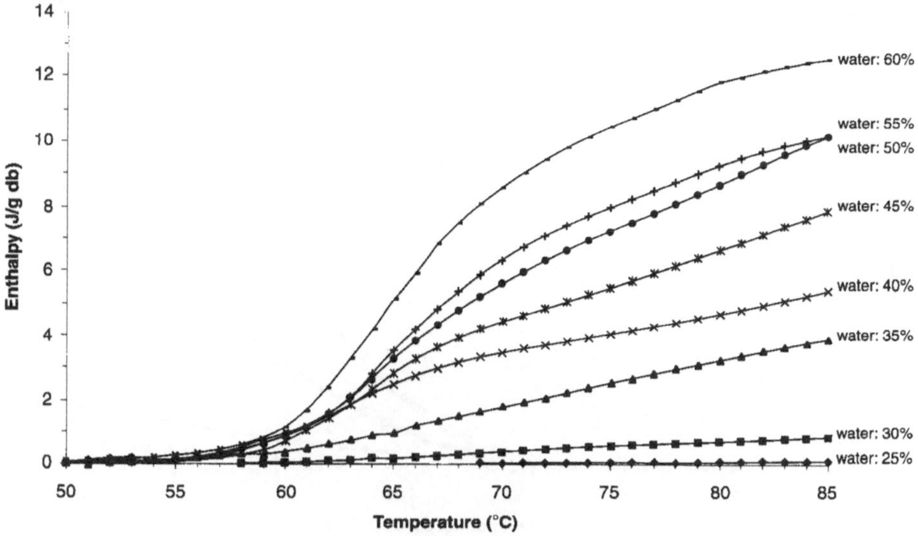

FIGURE 8 Evolution of melting enthalpy as a function of temperature for wheat starch preparations (moisture contents expressed on a wet basis). The enthalpy is calculated by integration of the area below the gelatinization peak from the onset of the peak to the selected temperature.

resulting from the progressive disappearance of the crystalline regions (i.e., the steric constraints controlling swelling) in starch and, thus, granule close-packing could be reached. The shifting of this event towards higher temperatures for the waxy corn starch is in agreement with DSC results,[6,12] showing that the onset of the crystallites' melting starts at a lower temperature for wheat than for corn starch (normal or waxy). It can thus be postulated that, upon heating, wheat starch starts swelling at a lower temperature than corn starch and the preparation reaches its ϕ_m at a lower temperature than the corn starch.

At the temperature of 85°C, the modulus is similar for wheat and waxy corn starches that have the same water content. For both starches, at this temperature the modulus decreases slightly as the initial water content increases. The values of the modulus for these closed packed suspensions suggest that the granular material is in the rubbery state. Heating to above the temperature where the maximum in G′ is observed provides energy to break down the residual crystalline structure of starch, causing G′ to drop down.[14] DSC results on waxy corn starch showed evidence of melting/dissociation of ordered zones in this range of temperature (Figures 8 and 9).

Influence of other ingredients

Adding sucrose to a 60% water-wheat starch preparation shifted starch melting observed with DSC and the increase in G′ observed with DMTA towards higher temperatures (Figures 10 and 11).[15]

Figure 12 compares the evolutions of the elastic modulus of gluten and wheat dough at the same water content, and of more hydrated wheat starch mixtures (with $\phi < \phi_m$ and $\phi > \phi_m$, respectively). This figure confirms that before heating, and when

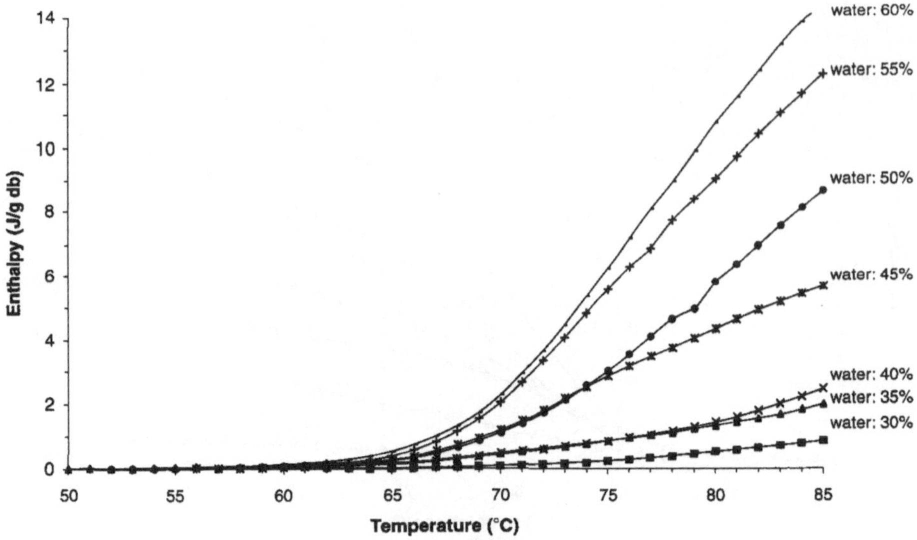

FIGURE 9 Evolution of melting enthalpy as a function of temperature for waxy corn starch preparations (moisture contents expressed on a wet basis). The enthalpy is calculated by integration of the area below the gelatinization peak from the onset of the peak to the selected temperature.

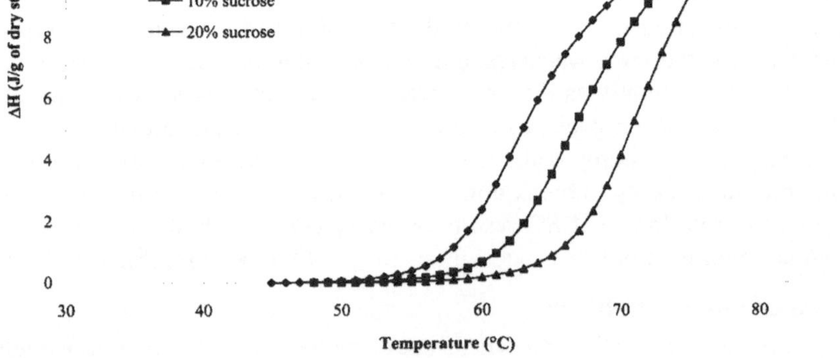

FIGURE 10 Effect of sucrose on partial melting enthalpy of wheat starch dispersions at 60% (w/w, wsb) moisture content.

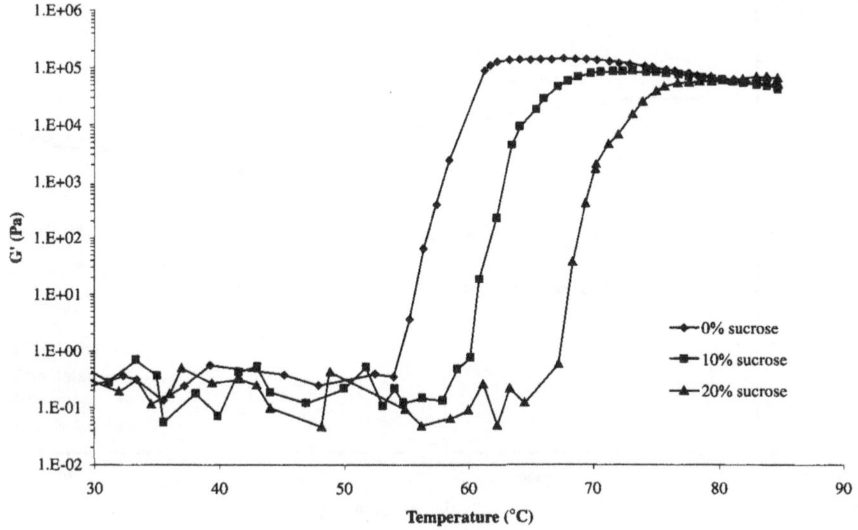

FIGURE 11 Effect of added sucrose on the storage modulus of 60% water content (w/w wsb) wheat starch dispersions during heating.

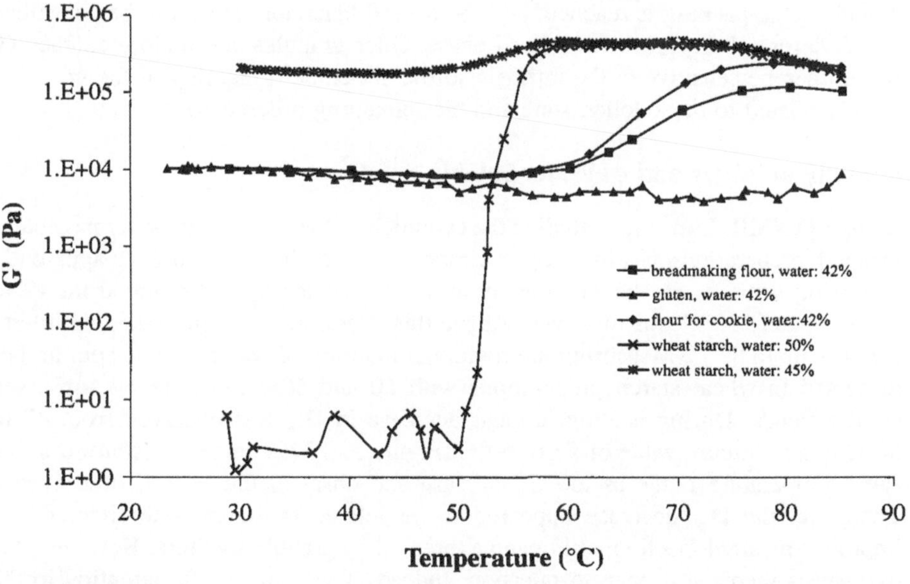

FIGURE 12 Storage modulus evolution during heating of wheat dough (42% water), gluten (42% water) and wheat starch (45 and 50% water).

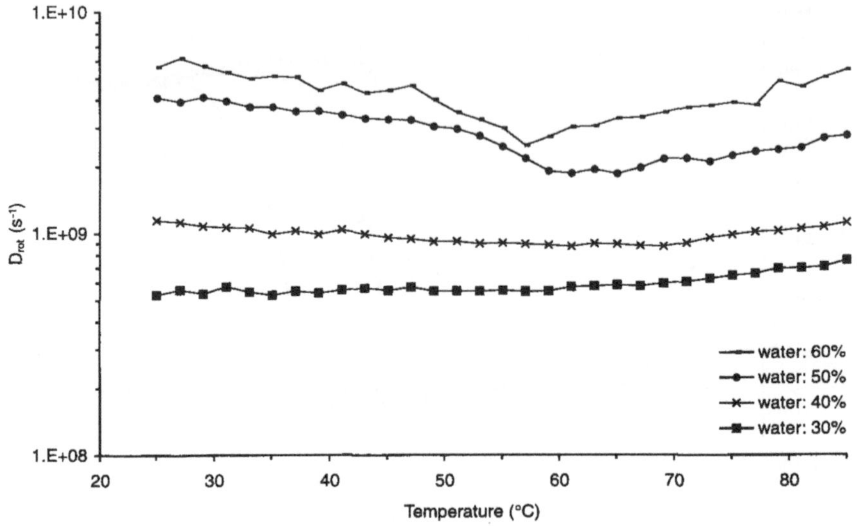

FIGURE 13 Rotational diffusion coefficient of spin probe (TEMPOL) dispersed in the aqueous phase of wheat starch dispersions at intermediate moisture contents (w/w, wb), as a function of temperature.

the volume fraction of starch is lower than its close-packing value, the continuous phase (i.e., the gluten network) controls the modulus of the mixture. After heating, granule close-packing is reached, and the overall behavior appears to be controlled by the deformability of the dispersed phase. Once granules are in close contact, G′ would then be sensitive to the intrinsic softness, i.e., deformability of the granules, closely related to the swollen state and the remaining ordered zones.

Mobility of Water and of Water-Soluble Solutes

Using ^{17}O NMR, Lim et al.[16] studied the evolution of the mobility of water molecules in starch preparations as a function of temperature. For 70% moisture, an apparently surprising decrease in the mobility of water molecules upon heating from 49 to approximately 53°C was observed. Above this, water mobility increased with temperature up to 87°C. Measuring the rotational mobility of water-soluble spin probes dispersed in wheat starch preparations with 60 and 50% moisture, we observed similar trends. During heating, a clear decrease in D_{rot} was observed from 47 to 50°C to a minimum value of 57 to 60°C (Figure 13). This decrease occurred in the same temperature range as the G′ increase and onset of the endothermic events. Therefore, the D_{rot} decrease appeared to be related to starch-water interactions improved by amylopectin melting and expressed by granule swelling. However, this hypothesis seems still open to question. Indeed, if we pay careful attention to the temperatures of onset of the different phenomena, it seems that the decrease in probe mobility would occur slightly before the onset of starch melting as revealed by DSC. Furthermore, waxy corn starch dispersions with 50 and 60% moisture, which also exhibit a similar melting process, did not show such decrease in D_{rot} (Figure 14).

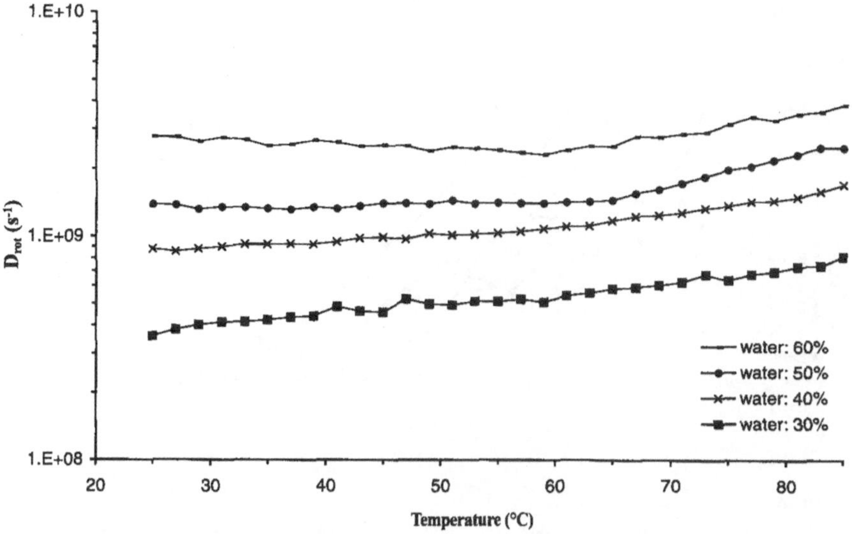

FIGURE 14 Rotational diffusion coefficient of spin probe (TEMPOL) dispersed in the aqueous phase of waxy corn starch dispersions at intermediate moisture contents (w/w, wb), as a function of temperature.

As for the melting process, sucrose addition to wheat starch with 60% water induced a shift of this decrease in D_{rot} at higher temperatures (Figure 15).

Only slight lineshape changes occurred during the thermal treatment of waxy corn starch compared to wheat starch. At 60 to 65°C, D_{rot} increased slightly for dispersions having 30 and 40% moisture contents. It was stable for 50% moisture content and slightly decreased for 60% moisture content. At 60-65 to 85°C, D_{rot} increased slightly for the lowest water contents of 30 and 40%, and more sharply for 50 and 60% moisture content.

The same ESR experiment was performed with wheat dough (42% water) with and without sucrose added. Similar trends for wheat starch with the same water content were observed, but the values of the rotational diffusion coefficient of the probes were significantly higher in dough than in starch preparations. Here, also, sucrose appeared to increase the thermal stability of starch-organized structures, partly because less water is available for starch hydration.

These ESR results showing the evolution of probe mobility during heating of starch preparations confirm that probes are sensitive to changes in starch structure and to the subsequent modifications of starch-water interactions. They also confirm the differences in wheat and waxy corn starch interactions with water. At room temperature, the presence of amylose in wheat starch could prevent the penetration of water towards some sites of amylopectin. Upon heating, amylose may start to dissolve and leach out of the granules, at least for the most hydrated samples. The subsequent opening up of new sites in domains that water molecules were not able to reach at lower temperatures may partly explain the decrease in D_{rot} observed for wheat starch in the 50 to 60°C temperature range.

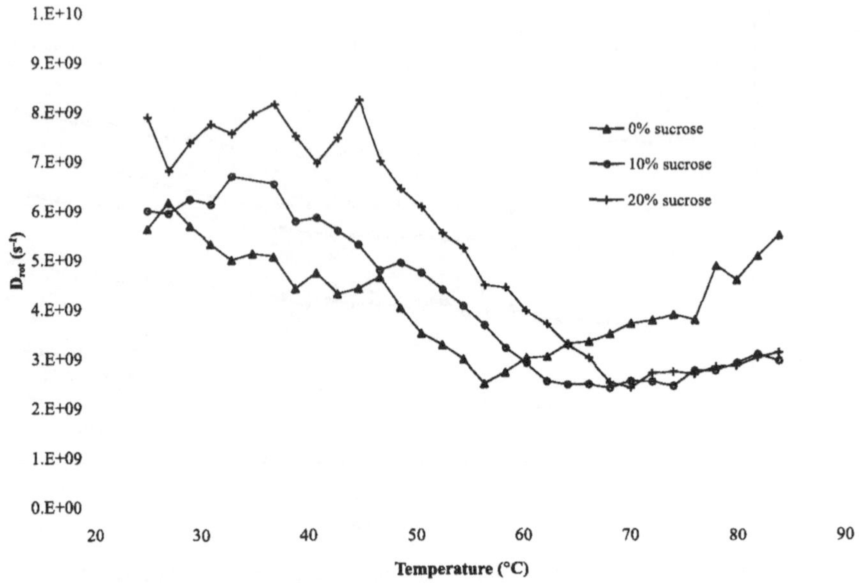

FIGURE 15 Effect of sucrose on the rotational diffusion coefficient of spin probes (TEMPOL) dispersed in the aqueous phase of 60% water content (w/w wsb) wheat starch dispersions during heating.

CONCLUSION

Two interdependent parameters appear to control the rheological properties of concentrated starch-based preparations just after cooking: deformability of the starch granules, and water content. The volume fraction occupied by the granules is also a determinant parameter but within our experimental conditions, granule close-packing is always reached at the end of the heating process, whatever the initial water content. Initial moisture content controls the extent of structural disorganization during cooking and the amount of plasticizer available for starch chain flexibility. The granule deformability will thus be higher for higher initial moisture contents.

Other ingredients present or added to a dough will mainly act as diluent for starch before cooking and as competitors for water during and after cooking. The comparison between wheat and waxy corn starch suggests a higher thermal stability and a stronger affinity for water in waxy corn starch, but no specific influence of amylose on the rheological behavior of starch-based preparations was observed.

ACKNOWLEDGMENTS

The study was conducted with financial support from the Commission of the European Communities, Agriculture and Fisheries (FAIR) specific RTD programme, CT-96-1085: Enhancement of Quality of Foods and Related Systems by Control of Molecular Mobility. It does not necessarily reflect its view and in no way anticipates the commission's future policy in this area. The authors greatly appreciate the contribution of undergraduate students Patrick Calboo, Karine Leroy, and Fabio Barban.

REFERENCES

1. Tester, R. F. and Morrison, W. R., Swelling and gelatinization of cereal starches. I. Effects of amylopectin, amylose, and lipids, *Cereal Chem.*, 67, 551, 1990.
2. Freed, J. H. and Fraenkel, J., Theory of line width in electron spin resonance spectra, *J. Chem. Phys.*, 39, 326, 1963.
3. Krieger, I. M., Rheology of monodisperse lattices, *Adv. Colloid Interface Sci.*, 3, 111, 1972.
4. Ferry, J. D., *Viscoelastic Properties of Polymers*, John Wiley & Sons, New York, 1980, 641.
5. Eilers, 1964, cited by Ferry, 1980.
6. Rolée, A. and Le Meste, M., Thermomechanical behavior of concentrated starch-water preparations, *Cereal Chem.*, 74, 581, 1997.
7. Zeleznak, K. J. and Hoseney, R. C., The glass transition in starch, *Cereal Chem.*, 64, 121, 1987.
8. Hoseney, R. C. and Rogers, D. E., Mechanism of sugar functionality in cookies, in *The Science of Cookie and Cracker Production*, Faridi, H., Ed., Chapman & Hall, New York, 1994, 203.
9. Li, S., Dickinson, E., and Chinachoti, P., Mobility of "unfreezable" and "freezable" water in waxy corn starch by $_2$H and $_1$H NMR, *J. Agric. Food Chem.*, 46, 62, 1998.
10. Li, S., Dickinson, E., and Chinachoti, P., Proton relaxation of starch and gluten by solid-state nuclear magnetic resonance spectroscopy, *Cereal Chem.*, 73, 736, 1996.
11. Biliaderis, C. G. and Vaughan, D. J., Electron spin resonance studies of starch-water-probe interactions, *Carbohydr. Polym.*, 7, 51, 1987.
12. Rolée, A. and Le Meste, M., Effect of moisture content on thermomechanical behavior of concentrated wheat starch-water preparations, *Cereal Chem.*, 76, 452, 1999.
13. Kowert, B. and Kivelson, D., ESR linewidths in solution. VIII. Two component diamagnetic solvents, *J. Chem. Phys.*, 64, 5206, 1976.
14. Lii, C. Y., Tsai, M. L., and Tseng, K. H., Effect of amylose content on the rheological property of rice starch, *Cereal Chem.*, 73, 415, 1996.
15. Chiotelli, E., Rolée, A., and Le Meste, M., Effect of sucrose on the thermomechanical behavior of concentrated wheat and waxy corn starch-water preparations, *J. Agric. Food Chem.*, 48, 1327, 2000.
16. Lim, H., Setsze, C. S., Paukstelis, J. V., and Sobczynska, D., ^{17}O nuclear magnetic resonance studies on wheat starch-sucrose-water interactions with increasing temperature, *Cereal Chem.*, 69, 382, 1992.

8 Bread Microstructure

Christopher G. Oates

CONTENTS

MOLECULAR STRUCTURE OF THE MAJOR COMPONENTS OF BREAD

Good bread-quality wheat flour is an optimum blend of starch (70 to 80%), proteins (8 to 18%), lipids (~2%), pentosans (~2%), enzymes and enzyme inhibitors, and other minor components.[1,2] Such flour when hydrated forms dough, the basis of bread. The properties of dough are made possible due to a unique set of characteristics inherent to wheat flour. Particularly important is the capability of wheat flour dough to retain, during expansion, gas produced during fermentation and baking. If this is achieved, a well-expanded loaf of bread with a light, even crumb will be the end–product.

0-8493-8790-6/01/$0.00+$.50

PROTEIN

Proteins constitute about 8 to 18% of wheat flour, yet despite a relatively minor presence, compared to starch (~80%), this component is technologically the most important.[3] The technology of wheat proteins is expressed through their rheological properties, and bread making performance and is a direct consequence of the structures formed in the intermediate (dough) and final (bread crumb) products. These structures are in turn related to protein quantity and quality. Despite the importance of protein quality, which is reflected by protein composition, this entity is difficult to define because of the extremely heterogeneous nature of the proteins. The first level of classification for this complex mixture is two groups, gluten and non-gluten forming proteins. The gluten proteins are, for the most part, important for bread structure.

Wheat Gluten Proteins

The importance of wheat flour is attributed mainly to the unique viscoelestic properties of the gluten proteins, a water insoluble complex.[4,5] The gluten-forming proteins, which represent 80 to 90% of the total proteins of wheat flour, are classified into two groups based on their extractability and lack of extractability in aqueous alcohol, correspondingly known as gliadins and glutenin.[6] Neither group consists of pure proteins, and overlap exists between the two. Nevertheless, distinction between the two groups is important, as each imparts different functional characteristics to a dough system.

- When hydrated, gliadins behave mainly as a viscous liquid and confer extensibility, allowing the dough to rise during fermentation. When isolated they are sticky.
- Glutens provide elasticity and strength, preventing the dough from being over-extended and collapsing during fermentation or baking.[5]

A more recent classification[7] is based on the biological, chemical, and genetic relationships of the component polypeptides in the gluten complex. In this classification, the gluten complex is described by three groups of proteins representing prolamines that are (1) sulphur poor, (2) sulphur rich, or (3) high molecular weight. Within the three categories, gliadins and glutanins are described.

All the gluten protein polypeptides have at least two or three distinct structural domains — a central repetitive domain flanked by non-repetitive C-terminal and N-terminal domains.[8] A detailed discussion of the molecular composition of glutenin proteins is given by Khatkar and Schofield.

Gliadins

These proteins are classified into four sub-categories (α, β, γ, ω), and together the categories contain a large number of polypeptides, between 49[9] and 60.[10] Two categories, α and β are structurally closely related, and for the sake of convenience are regarded as a single group known as the α-type gliadins. Similarities shared by the α- and γ-gliadins include a high proportion of sulphur amino acids, and relatively few proline, glutamine, and phenylamine residues compared to the ω gliadins. These two groups are smaller than the ω gliadins, with molecular weights between 30 and 45,000. In contrast, the ω-gliadins are large (m_rs 44 to 88,000), are rich in gluatamine, proline, and phenylalanine (these three residues constituting 80% of the total), but contain little or no sulphur amino acids.

Glutenins

This highly heterogeneous collection of polypeptides forms multi-chained structures varying in molecular weight from 40,000 to several million. On reduction of the intermolecular disulphide bonds, two types of subunits are formed, known as the high molecular weight (HMW) subunit and low molecular weight (LMW) subunit. The low molecular weight subunit is subcategorized into two groups (B subunits 42 to 51,000 and C subunits 30 to 40,000). The high HMW subunits are larger, with weights of 80 to 160,000 (based on SDS) or 63 to 88,000 (based on amino acid sequence), and may be further classified into x-type (M_rs 83 to 88,000) and y-type (M_rs 67 to 74,000).[3] The HMW subunits constitute the main proportion of the subunits of total glutenin.

A model for polymeric glutenin is therefore described in which the HMW glutenin subunits are linked at their ends by disulphide bonds. The cysteine residues are mostly located on the ends of the subnits. The presence of one such structure, known as the glutenin macroploymer (GMP), composed of HMW and LMW subunits, is related to bread-making quality parameters.[11]

The high molecular-weight glutenin polymers exhibit a strong tendency to aggregate, reflecting polypeptide chain composition. The main attributes for this structure-forming behavior are a large potential for interpolypeptide chain disulphide bonds, considerable hydrogen bonding (due to an unusually high amount of glutamine), and potential for polar bonding of the many polar side chains and low ionic character of the gluten proteins. The low charge density of the gluten proteins, due to their low level of basic amino acids, prevents mutual charge repulsion between the wheat gluten proteins, and as a result fosters association by non-covalent interactions. This behavior forms the basis for structure formation.

Disulphide bonds are the principal covalent bonds within and between gluten polypeptides, and are of considerable importance in the structure-forming capacity of this group of proteins. Non-covalent interactions include ionic, hydrogen, and Van der Waals interaction. These bonds are generally weaker, but are important in the structure formation process.[12]

Gluten Proteins and Bread Making Quality

The presence or absence of certain HMW subunits in the polymeric glutenin are correlated with bread-making quality and are regarded as influencing the variation in bread-making quality among different wheat cultivars. As many as 20 HMW subunits have been identified, each cultivar containing 3 to 5 HMWsubunits. In terms of bread making quality, not all subunits are alike. Some are known to be better;

namely subunit 1, 2, 5 + 10 or 7 + 9.[13] The amount and position of cystein residues is an important determinant.

Starch

Macromolecules

Starch is a mix of two distinct polysaccharide fractions — amylose (27 to 31%) and amylopectin (69 to 73%). Both are composed of glucose but differ in size and shape. Amylopectin the larger fraction, (10^7 to 10^9 Da), is highly branched. Five percent of its structure is alpha 1,6 branches.[14] The chains are assembled in a cluster structure based on a model described by Robin.[15] Amylose is the smaller fraction (10^5 to 10^6 Da; DP 500 to 5,000), and posses very few branches, 9 to 20 per molecule, with chain lengths 4 to >100 glucose units.

Starch Granule

Native wheat starch exists in granules of defined shape and size. The granules of wheat are contained in two size populations, lenticular A granules, 15 to 35 μm, and smaller polyhedral B granules, smaller than 10 μm. The smaller B granules account for 33 to 50% of the total weight of wheat endosperm starch. The structure of the granules is semicrystalline and thought to encompass various levels of complexity.[16] The first level is cluster arrangement of amylopectin branches. This arrangement describes a structure characterized by alternating regions of ordered, tightly packed parallel glucan chains, and less ordered regions composed predominantly of branch points. A unit of amylopectin cluster is thus regarded as composed of an amorphous portion containing most of the tightly spaced branches (amorphous lamella) and a thin crystalline region, ~5 nm (crystalline lamella), containing the parallel glucans.[17] The size of each cluster is highly conserved at 9 nm.[18]

Native wheat starch granules contain about 30% crystallite material and yield X-ray diffraction patterns corresponding to A polymorphs. Crystallinity occurs within the ordered arrays of amylopectin, and is created by intertwining of chains with a linear length greater than ten glucose units, to form double helices. Crystallization, or double helix formation, can occur between adjacent branches in the same amylopectin branch cluster or between adjacent clusters in three dimensions.

Granules are inert, structurally stable entities for which the degree of stability cannot be explained by crystallinity alone — granules contain too little crystalline material. Further levels of structural complexity are invoked to devise models that explain this apparent anomaly and describe the manner in which amorphous and crystalline domains are aligned.[16] However, starch structure remains an enigma, as, unfortunately, much of the information required to generate a complete model is still lacking.[19] The most promising model describes the structure of a potato starch granule in which crystalline domains form continuous networks of left-handed helices.[19] Voids inside the super helices are assumed empty and are roughly 8 nm wide, and the interpenetrating super helices are assumed to form the skeleton on which the granule is developed. Considering the relatively minor variation in size, shape, and properties of starch granules, and the fact that nature conserves structural designs, it is not unreasonable to assume that this model will be similar for other granular starches.[18]

The relationship between amylose and amylopectin is not completely understood, though the amylose fraction is assumed to exist in the granule as an entity, separated from the amylopectin fraction. This affords amylose the ability to leach out from the granule. Assigning amylose a precise location within the granule evidently is not easy. Amylose has been located in bundles between amylopectin clusters or randomly interspersed among clusters in both the amorphous and crystalline regions.[20] Demonstrated by a series of crosslinking studies, amylose molecules are thought to be present in the granule as individual molecules, randomly interspersed among amylopectin and in close proximity in both the crystalline and amorphous phases. Location of amylose with respect to the amorphous — crystalline zones in wheat starch is mainly in the amorphous region. Size is an important criteria.

Despite its limited role in crystal formation, amylose can influence the arrangement of double helices in unit cells by interfering with the packing density of amylopectin chains. The mechanism is not fully understood, but is assumed to result from helix formation between amylose and amylopectin chains or via amylose-induced disruption within the amorphous layer.[18]

Granule Surface

Little information is available about molecular composition or arrangement at the starch granule surface. The basis of current models was first described by Lineback's[14] "hairy billiard ball". In this model, the granule surface is not smooth but characterized by protruding chains. Later modification by Stark and Lynn[21] describes a granule whose surface is characterized by ends of amylose chains and protruding amylopectin clusters, which are thought to be the start of the next growth layer. These immature amylopectin molecules could be the source of intermediate material often reported as the third starch fraction. Newly incompletely synthesized islands of amylopectin are loosely attached to the granule, and provide specific points for amylase attack, the intensity of which is tempered by smoothing of the surface as amylose molecules fill the gaps.[22] The granule surface is relatively impenetrable to large molecules such as amylases, due to tight packing of amylopectin chains.

The structural attributes for granule porosity account for passage of small molecules into the granule. Entry of hydrolyzing enzymes and other large molecules into the interior of starch granules is restricted, and only possible via pores or channels. Surface pores on granules of corn, sorghum, and millet are suggested to be openings to channels that penetrate in a roughly radial direction through the granule.[22,23] Transmission electron microscopy of starch granule sections has shown the presence of openings in a number of granules 0.1 to 0.3 μm in diameter, compared with 0.07 to 0.1 μm diameter for interior channels.

Presence of pores, especially if extensive, would result in a macroporous structure whose available surface area is much greater than the boundary surface area.[24] Surface openings and interior channels are accommodated in the model of Oostergetel and van Bruggen[18] by a channel running through the center of the superhelical structures.

Holes inside starch granules, located at the center of the maltese cross (the characteristic dark cross seen on starch granules under polarized light where the hilum is centered), are believed to exist and are demonstrated in wheat starch granules.[25] It is not clear if surface or interior holes are formed naturally or are merely an artifact of drying during processing or sample preparation. Pressures imposed on the granule by removal of water during drying could conceivably generate a hole at the weakest point (the hilum).

Modification of Starch Structure by Processing Conditions

Integrity of starch granules is essential for their optimal performance in bread.[26] Mechanical, chemical, or biochemical disruptions of the native granule have adverse effects on the bread-forming properties of wheat starch.[26] The effect of excessive mechanical starch damage is further aggravated by abnormally high levels of alpha amylase associated with sprouted grain.

Controlled use of enzymes is practiced in baking to improve the quality of baked bread. Alpha amylase is used in baking to make the starch granule surface more porous, thereby allowing more amylose to leach out of the granule.

PROGRESSION OF BAKING CHANGES

Wheat Bread Dough

Changes during Mixing

Dough mixing is the most critical step in the bread-making process, and is responsible for a final structure involving complex interactions between the wheat flour proteins. Mixing is the process whereby ingredients are blended into a quasi-homogenous mixture within which a matrix of gluten proteins develops and air is trapped.[27]

A combination of physiochemical changes to the dough components occur, producing a final product. Flour proteins undergo the most important transformation. The formation of a gluten matrix is a process of aggregation and mechanical transformation of materials that exist in the pre-formed state within the endosperm cells. The most critical transformation is formation of the gluten matrix, a film-like structure, which starts at the branching points of protein strands and results in a layer-like arrangement in dough and gluten.[28] Gliadin and glutenin molecules associate (matrix formation) or dissociate (matrix disruption) according to the mixing treatment. The two processes occur during mixing, and optimally developed dough requires the correct balance between them. Protein aggregation is promoted by hydration, mixing, and other favorable factors. Disruption occurs due to either aggregate breakdown induced by over-mixing, or factors such as poor wheat quality or excessive polypeptide association caused by low mixing speeds. Optimum bread quality, resulting when a balance between matrix formation and disruption is attained, is the point of optimum dough development.[29] Protein matrix formation seems to involve the glutenin macro polymer, which during mixing is partly depolymerized, causing the liberation of SDS-extractable polymers. These polymers will re-polymerize during resting to form the glutenin macro polymer of dough.[11]

Underdevelopment, the result of either over- or under-mixing, promotes conversion of the continuous membranous structure of gluten into discontinuous fibrillar and globular structures in which starch and proteins are unevenly distributed and compact protein masses are stretched out into sheets. Both protein classes, but especially glutenins, show a marked tendency to associate during underdevelopment to form aggregates. Underdeveloped dough is characterized by structures that appear to be less fibrous and with larger diameter gliadin fibrils of about 5 micron. The glutenin fraction is predominantly sheet-like in structure, 5 to 50 micron.[29]

Dough Structure

Two models for the gluten matrix structure are usually quoted.[30] The older one defines a gluten matrix composed of long strands of glutenin molecules forming networks by associating at intervals, but which for the most part are separated by gliadin molecules. The second model defines specific associations between protein molecules from glutenin and gliadin fractions, leading to the creation of linear microfibrillar structures. The microfibrils (5 to 8nm) subsequently associate into macrofibils (up to 500nm), which in turn associate into bundles and sheets possessing the elastic and viscous properties of gluten. The gliadin and glutenin fractions of optimally mixed dough are therefore arranged in a highly fibrous structure.[29] In this model, secondary bonding forces, namely hydrogen, ionic, and hydrophobic bonds are important.

Optimally developed dough is one in which the continuous and interconnected gluten matrix surrounds most of the starch granules. Such dough also contains occluded gas cells whose diameters are typically 10 to 100 μm.[31] A precise determination of the size distribution of gas cells is difficult because microscopic examination underestimates the number of small gas cells, and those with a diameter smaller than the thickness of the section may not be observed.[27] The number and size of gas cells is influenced by mixing conditions and has a significant impact on the final bread character.

Flour Quality

Protein content and quality are critical factors that determine the ability of dough to develop to the desired extent. In good bread-making flours, underdevelopment is characterized by a dough structure in which the continuous membranous structure of gluten is converted into a discontinuous fibrillar and globular structure. For weak flours, irrespective of mixing treatment, a discontinuous gluten network is formed in which starch granules appear to be only partially covered by the gluten membrane. These doughs are characterized by structures that appear to be ruptured gluten membranes with many open areas. The gliadin and glutenin fractions of optimally developed dough of weaker flour also contain numerous spherical particles (2 to 4 microns in diameter). The gliadins do not form the fine (2 to 3 micron diameter) fibrils present in the glutenin. In contrast, gliadin appears as strands, intermixed with the numerous spherical particles. The small particles of glutenin seem to be linked to the fibrils. Underdevelopment produces a significant change in the gliadin and glutenin microstructure. Both fractions contain larger (6 to 12 micron) spherical globules than the corresponding spherical fraction from optimally developed dough.

Starch-Gluten-Water Relations

Association between starch granules and gluten proteins is affected by the quality of flour. Gluten proteins of poor quality flour interact more strongly with starch granules than starch granules and gluten proteins of good quality flour. The starch granules seem to act as inert filler, and interact with gluten proteins to effectively form crosslinks. The interactions apparently decrease the flow properties of poor quality dough. Differences in the starch-gluten interaction seem to be related to the proteins of good or poor quality flour, and not to differences in the starch. The mechanisms are not known.

Fermentation

Major changes taking place during fermentation lead to further development of the dough structure and formation of gas cells. Yeast fermentation generates carbon dioxide, and the dough expands due to increasing pressure in the gas cells.

Yeast cells contained in the aqueous phase within a dough's matrix ferment sugar to produce carbon dioxide. At some point, carbon dioxide will saturate the aqueous phase, allowing newly produced carbon dioxide to vaporize into the atmosphere or into pre-existing air cells.[32] The already partially developed dough network presents a barrier to the passage of carbon dioxide. Consequently, most of the carbon dioxide diffuses into air cells that are in closer proximity to the site of fermentation. With the expansion of air cells, a result of the excess pressure due to carbon dioxide uptake, the dough expands. Gas cell growth depends, in part, on their initial size. Greater pressure is needed to expand a small gas cell, so the smallest cells may never expand. Such small cells have been seen between large bubbles in a dough matrix.[33]

Flour quality affects the ability of dough to retain gas. The tight interaction between starch granules and gluten that characterize flours of low quality cause reduced viscous flow behavior and consequently less gas retention. Gas cells within such a system will not be able to expand as carbon dioxide flows into them. Consequently, the increasing internal pressure prevents carbon dioxide from diffusing into the air cells and results in its loss from the dough system.[34] The dough, therefore, does not expand. A further negative influence on dough structure when poor quality flour is used is that air cell sizes are not homogenously distributed. Pressure changes in an inverse proportion to bubble size.[34] As a result, the composition of gas cells changes into two distinct populations of large and small cells. Differences between the two size populations increase over time, resulting in a coarse dough structure. In contrast, the more viscoelastic dough of good quality flour allows gas cells to expand and creates a structure containing large, uniform gas cells.

Gas cell stabilization and retention determine crumb structure and volume in wheat bread. Sandwich breads are characterized by a large number of small bubbles of uniform size, whereas the size of bubbles in baguettes is more random.[27] Dough containing a large number of small bubbles will be more stable to freezing than dough containing bubbles of less uniform size, or if the walls surrounding larger bubbles are too thin.[35,36]

The spherical gas cells are lost at some point, either during late fermentation or early in baking. At this point, the dough structure becomes foam-like and is characterized by elongated polyhedral cells. This process is continuous, and the gas cell surface is thought to undergo major change.[37] The starch–gluten matrix at the gas-cell interface gives way to a stabilized liquid film, which eventually ruptures in baking. Gas cells are discrete during the first stage of fermentation, but as they expand, discontinuities develop in the starch-gluten matrix. The discontinuous nature of this structure forms over time. It is not evident in dough immediately after mixing, but can be seen after 15-minute proofing, and will increase in surface area thereafter. The discontinuities give the appearance of interconnections between gas cells and represent up to about 80% of the total dough volume at the end of proofing. The holes in the gas cell walls vary considerably in size, to 1mm or greater. Most are 50 to 500 μm. However, on close inspection gas cells are seen to be intact, partly separated by a fragile membrane or liquid film.[38] This film has the capacity to expand in response to the internal pressure generated by gas production. The discontinuous nature of the matrix can be seen clearly by SEM, and indirect evidence for this film of water provided by cryoSEM shows ice crystals forming on freezing, in the liquid phase of the dough covering the gas cell surface. Sublimation of this surface ice exposes a ruptured starch-gluten matrix.

Dough structure is more developed after fermentation than in the fresh state, and is thinner, finer, and better distributed. In addition, small gas cells will have developed that are distributed throughout the structure. The sheet-like protein structure of fresh dough is converted to a more complete network. Additional structure is provided by the gluing together, by gluten proteins, of strings of small starch granules into a strung-out matrix. The contribution of these starch structures to that of the dough is not known, but may be significant. Granules below 10 μm diameter account for 33 to 50% of the total weight of wheat endosperm. Strong interactions between protein and starch exist in fermented dough,[39] however the large starch granules may provide little or no direct contribution to the structure formation other than swelling and interaction with the gluten network.

BAKING

During the baking process, a number of heat-triggered changes to the dough structure[40] lead to transformation of dough into a sponge-like structure, the bread-crumb. The changes determine volume of the final bread loaf. Heating during baking causes expansion of gas cells and further stretching and thinning of the protein sheets. Starch granules that are entirely enrobed by protein gelatinize and flexibly fit around air cells, thereby reinforcing the gluten system. Because of limited availability of water during gelatinization, swelling is limited. Consequently, many granules remain intact and identifiable, despite their highly deformed irregular shape. The extent of gelatinization depends on the available water and temperature during baking.[41-43] Starch gelatinization also results in dewatering of the gluten phase,[39] causing dense granule appearance of the protein through localized condensation of the gluten matrix. The extent of this is dependent on temperature, humidity, and duration of baking.

Within the swollen granules, both amylose and amylopectin are present and are in an amorphous state directly after gelatinization. A small amount of amylose leaches out of the granule and forms a thin layer of amylose gel between the granules. Sampling from the crust through the breadcrumb reveals differences in the integrity of starch granules. Granules in the crust retain their shape, size, and appearance.[26] By the center of the crumb, most of the large granules are gelatinized, and small ones remain unaltered in appearance. Differences in degree of starch modification seem to be related to water availability for starch gelatinization. More water is available in the area near the crumb, and a significant amount is evaporated from the area immediately below the outside of the crust. The contribution of small granules in baked bread is relatively minor, while large granules swell and interact with the thin protein strands. The final structure of baked bread, therefore, involves interactions of denatured gluten, swollen starch (mainly large granules), and small starch granules that are strung together.

Another major structural change during heating of wheat dough is expansion of gas cells into an open network of pores.[44] As the lamella between gas cells ruptures, the cells become interconnected, forming a sponge structure. Stabilization of the gas-liquid interface, by migration of lipid crystals as the gas cell expanded during fermentation, increases early in baking when the lipid crystals melt. This allows the gas bubbles to grow without rupturing, giving the bread a large volume.[27] The porous structure of mainly open polyhedral cells is best described as a solid elastic sponge. Evidence suggests that very small gas cells maintain their integrity at this stage.

Flexibility of the gluten matrix in allowing gas cells to expand during the early baking stage is important. A rigid gluten structure that resists expansion of the gas cells will lose most of its carbon dioxide during the early stages of baking.[34] If gas cells freely expand, carbon dioxide remains contained in the dough, producing a large loaf with good oven spring.

At some point, structural changes take place which cause the bread to set. The gluten proteins undergo a number of structural changes originally thought to be associated with denaturation. More recent evidence suggests that gluten, under the conditions present in dough, is a random structure.[45,46] The changes to gluten are thought to be due to cross-linking of the protein[45,47] by disulphide interchange reactions. Crosslinking is assumed responsible for the setting of bread.[48] Gluten proteins also depolymerize on heating,[49] thus contributing to the final structure.

Crumb Grain Structure

The final porous breadcrumb structure reflects the changes during baking and subsequent cooling. This structure has been the subject of many studies.[27,39,50-56] In the breadcrumb, solid material is distributed in lamellae, and beams composed mainly of a bicontinuous system of interwoven gluten and gelatinized starch strands. Owing to the limited water present in bread dough, many of the starch granules are only partly gelatinized and swollen. Those that have gelatinized will have created a fibrous structure. A proportion of amylose that has leached out of the granule into the intergranular spaces will have undergone a number of phase changes that ultimately provide structural elements to the crumb. These elements prevent the crumb from collapsing.[57] Both amylose and amylopectin are present in the swollen granules, and it is on this level that structural changes occur during storage. Protein is converted to even sheets of dense particles that contain small inclusions. The protein strands are thin, and small vacuoles are evident. The granular ultrastructure of protein appears denser than that of starch. The connection between protein and starch is firm, and the protein envelops the starch. Free water is not present, having been taken up by starch. The walls of the remaining gas cells are very thin.

The coherence and continuity of the protein matrix is often weakened by small particles and disrupted by large particles such as bran,[26] dead yeast cells, and insoluble cell wall fragments. Hemicellulases are used to decrease the number of such fragments.[27] Bran particles can also effect gas cell development by causing distortion, restricting gas cells or forcing them to expand in a particular way.[58]

The structure of the crust is different from that of the crumb. It is a more porous system containing relatively large vacuoles. Starch granules are present inside the vacuoles, and these are less expanded.

REFERENCES

1. MacRitchie, F., Baking quality of wheat flours, *Adv. Food Res.*, 29, 201, 1984.
2. Pomeranz, Y., Composition and functionality of wheat flour components, in *Wheat Chemistry and Technology*, Vol. 2, Pomeranz, Y., Ed., American Association of Cereal Chemists, St. Paul, 1988, 97.
3. Khatkar, B.S. and Schofield, J.D., Molecular and physico-chemical basis of breadmaking — properties of wheat gluten proteins: a critical appraisal, *J. Food Sci. Technol.*, 34, 85, 1997.
4. Tatham, A.S., Shewry, P.R., and Miflin, B.J., Wheat gluten elasticity: a molecular basis to elastin, *FEBS Letts.*, 177, 205, 1984.
5. Khatkar, B.S., Bell, A.E., and Schofield, J.D., The dynamic rheological properties of glutens and gluten sub-fractions from wheats of good and poor breadmaking quality, *J. Cereal Sci.*, 22, 29, 1995.
6. Osborne, T.B., *The Proteins of the Wheat Kernel*, Carnegie Institute, Washington, D.C., 1907.
7. Shewry, P.R., Tatham, A.S., Forde, J., Kreis, M., and Miflin, B.J., The classification and nomenclature of wheat gluten proteins: a reassessment, *J. Cereal Sci.*, 4, 97, 1986.
8. Shewry, P.R., Tatham, A.S., Barro, F., Barcelo, P., and Lazzeri, P., Biotechnology of breadmaking: unraveling and manipulating the multi-protein gluten complex, *Bio/Technology*, 13, 1185, 1995.
9. Branlard, G. and Dardevet, M., Diversity of grain proteins and bread wheat quality. I. Correlation between gliadin bands and flour quality characteristics, *J. Cereal Sci.*, 3, 329, 1985.
10. Campbell, W.P., Wrigley, C.W., Cressey, P.J., and Slack, C.R., Statistical correlations between quality attributes and grain protein composition for 71 hexaploid wheat used as breeding parents, *Cereal Chem.*, 64, 293, 1987.

11. Weegels, P.L., Hamer, R.J., and Schofield, J.D., Depolymerization and re-polymerization of wheat glutenin during dough processing. II. Changes in composition, *J. Cereal Sci.,* 25, 155, 1997.
12. Belitz, H.D., Kieffer, R., Seilmeier, W., and Wieser, H., Structure and function of gluten proteins, *Cereal Chem.,* 63, 336, 1986.
13. Payne, P.I., Nightingale, M.A., Krattiger, A.F., and Holt, L.M., The relationship between high M glutenin subunit composition and the bread making quality of British grown wheat varieties, *J. Sci. Food Agric.,* 40, 51, 1987.
14. Lineback, D.R., Current concepts of starch structure and its impact on properties, *Dempun Kagaku,* 33, 80, 1986.
15. Robin, J.P., Mercier, C., Duprat, F., Charbonniere, R., and Guilbot, A., Lintnerized starches. Chromatographic and enzymatic studies of insoluble residues from acid hydrolysis of various cereal starches, particularly waxy maize starch, *Starch,* 27, 36, 1975.
16. Oates, C.G., Towards an understanding of starch granule structure and hydrolysis, *Trends Food Sci. Technol.,* 8, 375, 1997.
17. Hizukuri, S., Towards an understanding of the fine structure of starch molecules, *Dempun Kagaku,* 40, 133, 1993.
18. Oostergetel, G.T. and van Bruggen, E., The crystalline domains in potato starch granules are arranged in a helical fashion, *Carbohydr. Polym.,* 21, 7, 1993.
19. Svensson, E. and Eliasson, A.C., Crystalline changes in native wheat and potato starches at intermediate water levels during gelatinization, *Carbohydr. Polym.,* 26, 171, 1995.
20. Jane, J., Xu, A., Radosavljevic, M., and Seib, P.A., Location of amylose in normal starch granules. I. Susceptibility of amylose and amylopectin to cross-linking reagents, *Cereal Chem.,* 69, 405, 1992.
21. Stark, J.R. and Lynn, A., Biochemistry of plant polysaccharides: starch granules large and small, *Biochem. Soc. Trans.,* 20, 7, 1999.
22. Fannon, J.E., Shull, J.M., and BeMiller, J.N., Interior channels of starch granules, *Cereal Chem.,* 70, 611, 1993.
23. Baldwin, P.M., Adler, J., Davies, M.C., and Melia, C.D., Holes in starch granules: confocal, SEM and light microscopy studies of starch granule structure, *Starch,* 46, 341, 1994.
24. Zhao, X.C. and Sharp, P.J., An improved 1-D SDS-PAGE method for the identification of three bread wheat 'waxy' proteins, *J. Cereal Sci.,* 23, 191, 1996.
25. Adler, J., Baldwin, P.M., and Melia, C.D., Starch damage. II. Types of damage in ball-milled potato starch observed by confocal microscopy, *Starch,* 47, 252, 1995.
26. Pomeranz, Y., Meyer, D., and Seibel, W., Wheat, wheat-rye and rye dough and bread studied by scanning electron microscopy, *Cereal Chem.,* 61, 53, 1984.
27. Autio, K. and Laurikainen, T., Relationships between flour/dough microstructure and dough handling and baking properties, *Trends Food Sci. Technol.,* 8, 181, 1997.
28. Amend, T. and Belitz, H.D., Gluten formation studied by the transmission electron microscope, *Z. Lebensm. Unters. Forsch.,* 191, 184, 1990.
29. Paredes, L.O. and Bushuk, W., Development and 'undevelopment' of wheat dough by mixing: microscopic structure and its relations to bread-making quality, *Cereal Chem.,* 60, 24, 1983.
30. Mecham, D.K., Wheat proteins — observations on research problems and progress, Part 1, *Food Technol. Aust.,* 32, 540, 1980.
31. Moss, R., Bread microstructure as affected by cysteine, potassium bromate and ascorbic acid, *Cereal Foods World,* 20, 289, 1975.

32. Hoseney, R.C., Gas retention in bread, *Cereal Foods World,* 29, 305, 1984.
33. Bruijne, D.W., de Loof, J., and Eulem, A., The rheological properties of bread dough and their relation to baking, in *Rheology of Food, Pharmaceutical and Biological Materials with General Rheology,* Carter, R.E., Ed., Elsevier, London, 1990, 269.
34. He, H. and Hoseney, R.C., Factors controlling gas retention in nonheated doughs, *Cereal Chem.,* 69, 1, 1992.
35. Rasanen, J., Haerkoenen, H., and Autio, K., Freeze-thaw stability of prefermented frozen doughs: the effect of flour quality and fermentation time, *Cereal Chem.,* 72, 637, 1995.
36. Rasanen, J., Laurikainen, T., and Autio, K., Fermentation stability and pore size distribution of frozen prefermented lean wheat dough, *Cereal Chem.,* 74, 56, 1997.
37. Gan, Z., Angold, R.E., Williams, M.R., Ellis, P.R., Vaughan, J.G., and Galliard, T., The microstructure and gas retention of bread dough, *J. Cereal Sci.,* 12, 15, 1990.
38. Gan, Z., Ellis, P.R., and Schofield, J.D., Mini review: gas cell stabilization and gas retention in heated bread dough, *J. Cereal Sci.,* 21, 215, 1995.
39. Fretzdorff, B., Bechtel, D.B., and Pomeranz, Y., Freeze-fracture ultrastructure of wheat flour ingredients, dough, and bread, *Cereal Chem.,* 59, 113, 1982.
40. Moore, W.R. and Hoseney, R.C., The effect of flour lipids on the expansion rate and volume of bread baked in a resistance oven, *Cereal Chem.,* 63, 172, 1986.
41. Yasunaga, T., Bushuk, W., and Irvine, G.N., Gelatinization of starch during baking, *Cereal Chem.,* 45, 269, 1968.
42. Blanshard, J.V., Starch granule structure and function: a physicochemical approach, *Crit. Rep. Appl. Chem.,* 13, 16, 1987.
43. Lineback, D.R. and Wongrikasem, F., Gelatinization of starch in baked products, *J. Food Sci.,* 145, 71, 1980.
44. Eliasson, A.C. and Larsson, K., Basic concepts of surface and colloid chemistry, in *Cereals in Breadmaking, A Molecular Colloidal Approach,* Eliasson, A.C. and Larsson, K., Eds., Marcel Dekker, New York, 1993,1.
45. Schofield, J.D., Bottomley, R.C., LeGrys, G.A., Timms, M.F., and Booth, M.R., Effect of heat on wheat gluten, in *Proceedings 2nd Workshop on Gluten Proteins,* Graveland, A. and Moonen, J.M.E., Eds., TNO, Wageningen, 1984, 81.
46. Eliasson, A.C., Gudmundsson, M., and Svensson, G., Thermal behaviour of wheat starch in flour — relation to flour quality, *Lebensm. Wiss. Technol.,* 28, 227, 1995.
47. Schofield, J.D., Bottomley, R.C., Timms, M.F., and Booth, M.R., The effect of heat on wheat gluten and the involvement of sulphydrl-disulphide interchange reactions, *J. Cereal Sci.,* 1, 241, 1983.
48. Bale, R. and Muller, H.G., Application of statistical theory of rubber elasticity to the effect of heat on wheat gluten, *J. Food Technol.,* 5, 295, 1970.
49. Hoseney, R.C. and Rogers, D.E., The formation and properties of wheat flour doughs, *Crit. Rev. Food Sci. and Nutr.,* 29, 73, 1990.
50. Moss, R., Dough miscrostructure as affected by the addition of cysteine, potassium bromate, and ascorbic acid, *Cereal Sci. Today,* 19, 557, 1974.
51. Marston, P.E. and Wannan, T.L., Bread baking — the transformation from dough to bread, *Bakers Dig.,* 50(4), 24, 1976.
52. Hoseney, R.C., Atwell, W.A., and Lineback, D.R., Scanning electron microscopy of starch isolated from baked products, *Cereal Foods World,* 22, 56, 1977.
53. Khoo, U., Christianson, D.D., and Inglett, G.E., Scanning and transmission microscopy of dough and bread, *Bakers Dig.,* 49(4), 24, 1975.
54. Bechtel, D.B., Pomeranz, Y., and Francisco, A., Breadmaking studied by light and transmission electron microscopy, *Cereal Chem.,* 55, 392, 1978.

55. Chabot, J.F., Hood, L.F., and Liboff, M., Effect of scanning electron microscopy preparation methods on the ultrastructure of white bread, *Cereal Chem.*, 56, 462, 1979.
56. Vassileva, R., Seibel, W., and Meyer, D., Baking parameters and bread quality. IV. Structural changes of starch in bread crumb during baking (Backparameter und brotqualitaet. IV. Strukturveraenderungen der staerke in der brotkrume beim backen), *Getreide,* 35, 303, 1981.
57. Blanshard, J.V., Structure and function of the starch granule, in *Chemistry and Physics of Baking: Materials, Processes and Products*, Blanshard, J.V., Frazier, P.J., and Galliard, T., Eds., The Royal Society of Chemistry, London, 1986, 1.
58. Gan, Z., Galliard, T., Ellis, P.R., Angold, R.E., and Vaughan, J.G., Effect of the outer bran layers on the loaf volume of wheat bread, *J. Cereal Sci.*, 15, 151, 1992.

9 Modeling the Kinetics of Starch Retrogradation

Imad A. Farhat and J.M.V Blanshard

CONTENTS

INTRODUCTION

It is now more than 70 years since Katz, using x-ray diffraction, established the relationship between the so-called staling (i.e., the changes on storage of textural and mouthfeel attributes of bread and other baked goods), and recrystallization of amorphous gelatinized starch. This recrystallization of converted starch, also referred to as starch retrogradation has received extensive attention due to its scientific and economic importance.[2-4]

The various steps in obtaining retrogradation rate constants that can be modeled within the framework of a polymer science approach are reviewed here, showing the effects of storage temperature, water, and possibly other plasticizing molecules on the kinetics of starch retrogradation in the concentrated system conditions contiguous to the glass-rubber transition region. Most of the reported work on starch retrogradation deals with dilute and semi-dilute systems, which are of little relevance

0-8493-8790-6/01/$0.00+$.50

to the baking industry, particularly in the case of stability of finished products where water content rarely exceeds one gram of water per gram of dry matter.

MODELING THE KINETICS OF RETROGRADATION

Crystallization Kinetics

During retrogradation, crystals can grow from different nucleation centers. The amount of crystallized material present at a given time is, therefore, a combined function of crystal growth rate and the density of nucleation, reflecting the rate of nucleation. A stretched exponential equation derived from the Avrami[5] equation could therefore be used to model the progress of starch retrogradation, with storage time for a given composition and storage temperature:

$$Y(t) = Y_\infty - (Y_\infty - Y_0)\exp[(-G.t)^n] \quad (1)$$

where Y(t) is a physical characteristic reflecting the amount of crystallized material (XRD crystallinity index, NMR relaxation rates, elastic modulus), Y_0 and Y_∞ are the values of Y at storage times t = 0 and t = ∞ respectively, G is the rate of retrogradation ($time^{-1}$), and n is an Avrami exponent.

All four parameters could be reliably[6] determined from experimental data by non-linear optimization algorithms, assuming that sufficient data points have been acquired over a wide range of storage times (Figure 1). This approach is preferred to an approach based on linearizing Equation 1, since this can only be done by assigning the last measured value of Y(t) to Y_∞ and discarding any Y(t) value that exceeds Y_∞ (Le Botlan and Desbois).[7]

Effect of Temperature on the Crystallization Kinetics: The Crystallization of Chain Folded-Polymers Approach

In the 1960s, Lauritzen and Hoffman developed a theory describing the kinetics of crystallization of linear synthetic polymers in which chain folding constituted the mean of growth of the crystalline lamellae.[8,9] This approach describes the dependence of kinetics of the crystallization on temperature in relation to glass-rubber transition temperature and undercooling.

The dependence of the crystallization rate on storage temperature T (crystallization temperature) is given by:

$$G(T) = G_0\exp\left[-\frac{U^*}{R(T - T_\infty)}\right]\exp\left[-\frac{K_g}{T\ \Delta T\ f}\right] \quad (2)$$

where T is crystallization temperature (K), U* the activation energy for the steady state repetition of the polymer chain ($J.mol^{-1}$), R the gas constant ($R = 8.314\ J.K.^{-1}.mol^{-1}$), ΔT the undercooling $\Delta T = T_m - T$, T_m the melting temperature, and K_g a constant.

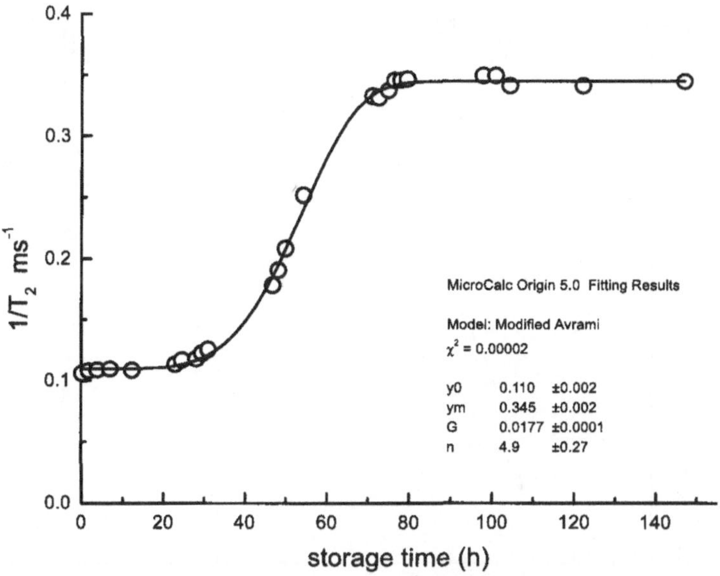

FIGURE 1 Kinetics of the isothermal retrogradation (40°C ± 0.1) of a 100:30 extruded waxy maize starch–water system as followed by the change in the ^{1}H NMR spin-spin relaxation rate measured with CPMG pulse sequence. The line depicts the best fit (obtained using the commercial software MicroCalc Origin) to the experimental data points, using Equation 1. The fitted values of the various parameters are also shown.

T_∞ is a hypothetical temperature at which viscous flow would cease. T_∞ is related to glass transition temperature T_g by the relationship $T_\infty = T_g - \delta T$.

The f factor accounts for change in the heat of fusion, with temperature $f = 2T/(T_m + T)$.

Application of Lauritzen-Hoffman Approach to Kinetics of Starch Retrogradation

The Lauritzen-Hoffman theory describing the growth of chain-folded polymer crystals may be applied to model the kinetics of starch retrogradation. The crystals formed during recrystallization of the amorphous (gelatinized) form of the highly branched amylopectin molecule have many similarities. Chain-folded polymeric crystals with the double helices of the A-chains form the crystalline lamellae, while the branching regions constitute the folds. This approach was first adopted by Marsh and Blanshard[10] for the kinetics of wheat starch gels. The analysis has since been improved[6] and extended to a wider range of compositions and storage conditions. This approach offers the opportunity to rationalize the effects of storage temperature and sample composition on the rate of retrogradation in terms of molecular mobility, and considers the impact of composition (water content) on the temperatures at which the glass-rubber and crystalline-amorphous transitions occur.

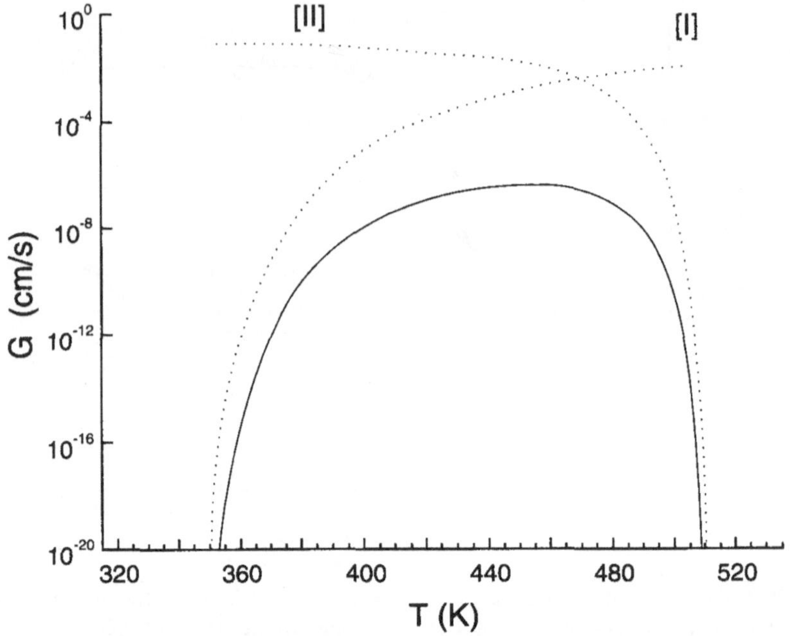

FIGURE 2 The fit obtained by Lauritzen and Hoffman,[9] using Equation 2 for experimental data relating to isotactic polystyrene. The parameters used in this simulation were $G_0 = 2.7$ 10^{-2} cm.s^{-1}, $U^* = 6530$ J.mole^{-1}, $T_\infty = 333.5$ K ($T_g - 30$), $Kg = 1.2 \times 10^5$ and $T_m = 515.2$ K. The dotted lines depict the behavior of the first and second terms of the equation.

In order to apply Equation 2 to the experimental retrogradation results, numeric values of the five parameters of the Lauritzen-Hoffman equation T_∞, T_m, U^*, K_g and G_0 are required. While T_g (and consequently T_∞) and T_m could be defined experimentally or estimated as described below, the values of U^*, K_g, and G_0 can be obtained by non-linear least squares minimization fitting of Equation 2 to experimental data sets.

Calculating the Parameters T_∞ and T_m

In the absence of measured values for T_g and T_m, it is possible to obtain calculated temperatures using the equations suggested by ten Brinke et al.[11] and Flory.[12] The merit of such models has been shown in predicting T_g[13-15] and T_m[16-19] over a wide range of water contents.

Calculation of T_∞

The calculation of T_∞ for different formulations involves the calculation of the T_g for each composition and then the application of the relationship $T_\infty = T_g - \delta T$. Lauritzen and Hoffman suggested that a value of $\delta T = 30$K was successful in describing the experimental results of several synthetic polymers.[9] Furthermore, Marsh and Blanshard[10] found that $\delta T = 30$K was suitable for modeling of retrogradation of 50% wheat starch gels. $T_\infty = T_g - 30$ was therefore adopted for this study.

The ten Brinke and Karasz equation derived from the Couchman-Karasz[20] equation describes the composition dependence of T_g based on a thermodynamic understanding of the glass-rubber transition:

$$Tg = \frac{W_s \Delta Cp_s Tg_S + W_p \Delta Cp_p Tg_p}{W_s \Delta Cp_s + W_p \Delta Cp_p} \tag{3}$$

where the subscripts $_s$ and $_p$ refer to the solvent (water) and the polymer (amylopectin) respectively, W is the weight fraction and ΔC_p is the difference in specific heat capacity between the liquid and glassy states at T_g with $Tg_p = 500$ K and $\Delta Cp_p = 0.41$ J g^{-1} K^{-1} for pregelatinized (amorphous) waxy maize starch, and, $Tg_s = 134$ K and $\Delta Cp_s = 1.94$ J g^{-1} K^{-1} for water.[13,14]

Calculation of T_m

The calculation of the melting temperature for different water contents is done according to the Flory equation, which suggests that the following relationship holds between the melting point of a polymer and the diluent concentration:

$$\frac{1}{T_m} - \frac{1}{T_m^0} = \frac{R}{\Delta H_u}\frac{V_u}{V_1}(v_1\chi_1 - v_1^2) \tag{4}$$

where ΔH_u is the change in enthalpy of fusion per repeating unit (glucosyl), V_u/V_1 is the ratio of molar volumes of the repeating unit in the chain to that of water, R is the gas constant, T_m (K) is the melting point of the polymer-diluent system, T_m^0 is the true melting point of the undiluted polymer, v_1 is the volume fraction of the diluent, and χ_1 is an interaction parameter.

In these calculations, values of $\Delta H_u = 25.4$ kJ.mole^{-1}, $\chi_1 = 0.5$, and $T_m^0 = 550$K were calculated by Farhat et al.[21] using the Flory approach and applying a non-linear least squares optimization to the experimental results reported by other workers. A value of 1.5 g.cm^{-3} was used for the density of starch.

EFFECTS OF STORAGE TEMPERATURE AND WATER CONTENT ON RETROGRADATION RATE

This section reports an example of the approach developed above to model the effect of temperature on the rate of starch retrogradation. In the final part of this section, the Lauritzen-Hoffman equation will be extended to predict the effect of water content on the rate of isothermal retrogradation. The study, carried out on waxy maize starch, is reported in more detail elsewhere.[6]

EXPERIMENTAL PROCEDURES

Non-expanded waxy maize starch (wms) water strips were prepared by twin screw extrusion (120°C). The water content was varied during extrusion, and samples containing 20 to 60% water (dry solid basis) were prepared, sealed, and stored at different temperatures.

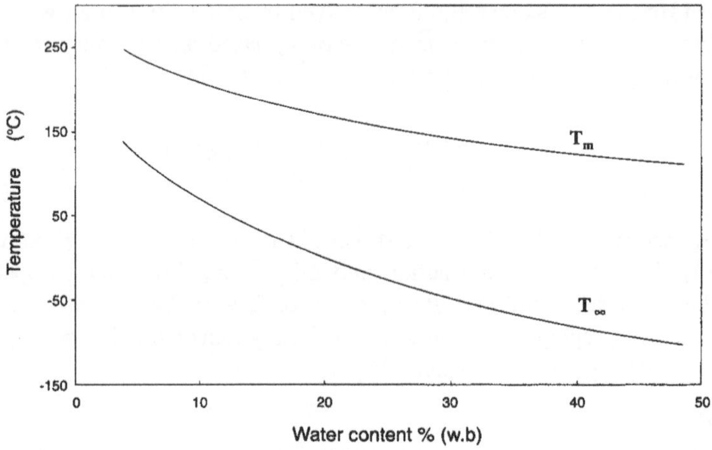

FIGURE 3 Variation of T_∞ and T_m with the composition of the sample as calculated using Equations 3 and 4.

Pulsed low-field ^1H NMR was used to follow the progress of retrogradation. In addition to providing information on change in molecular mobility of the various components throughout the recrystallization process, this technique offers several practical advantages, such as limited moisture loss during measurement and reliable temperature control. Due to its noninvasive nature, the same sample could be studied. Finally, NMR enables information about changes in the bulk of the sample monitored, not only the surface as in FTIR or XRD. The relatively large amount of sample studied is advantageous when the heterogeneity of the sample is inherent its nature (for example, wholemeal breads). NMR measurements were performed using a 20 MHz Bruker PC120 Minispec operating at 40°C ± 0.1. The spin-spin relaxation times (T_2) were obtained by fitting a single exponential to the CPMG decay, recorded with a 90 to 180° pulse spacing of 262μs.

Effect of Storage Temperature on Retrogradation Rate

The rate of starch retrogradation showed a typical bell-shaped dependence on the storage temperature (Figure 4). This behavior is in agreement the findings of other workers on similar systems[6,10,22] and the general theory of crystallization, where the effect of temperature on the rate of crystallization is the result of its effects on the nucleation and crystal growth steps.

If more water is incorporated in the system, the retrogradation rate versus temperature is anticipated to shift to lower temperatures due to the role of water in plasticizing the polymer and consequently decreasing its T_g (and therefore T_∞), as well as decreasing the temperature at which the polymer crystallites would melt. This is discussed in detail elsewhere.[6]

For bread with moisture content of approximately 40% (wet basis), the maximum of the retrogradation curve was found by Marsh and Blanshard (unpublished results) to occur at a temperature of approximately 4°C.

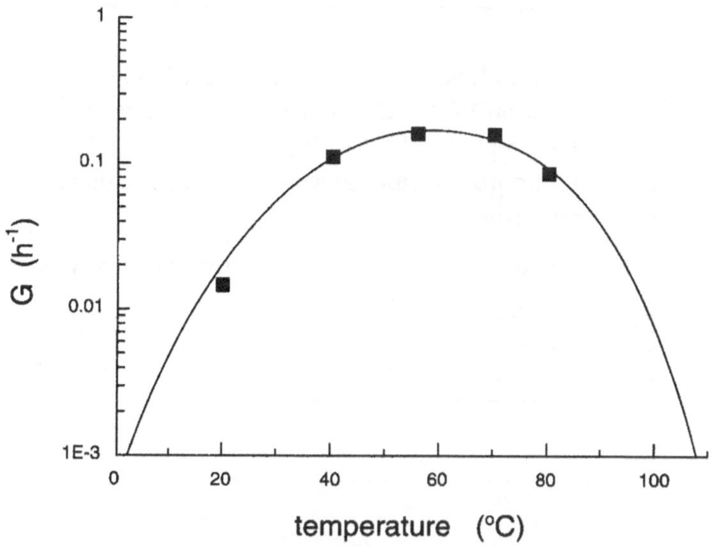

FIGURE 4 Effect of storage temperature on the rate of retrogradation of a wms system containing 30% water (wet basis). The line was obtained by fitting Equation 2 to the experimental data (adapted from Farhat et al.).[6]

Extending the Approach to Effect of Water Content on the Rate of Isothermal Retrogradation

Results similar to those shown in Figure 4 were obtained for different water contents.[6] The relevant T_g and T_m values were calculated using Equations 3 and 4, and the values of G_0, U*, and K_g were obtained by fitting Equation 2 to the experimental data. The parameters $\log G_0$, U*, and K_g decreased in a linear manner with increased water content. These dependencies are summarized in Table 1. The intercept values for the pure amylopectin of U* and K_g are of the same order of magnitude as those obtained by Lauritzen and Hoffman for linear synthetic polymers isotactic polystyrene and nylon-6. The fact that U* for amylopectin is approximately 30% greater than for these two synthetic polymers is probably due to its highly branched structure and extremely high molecular weight. This is consistent with the much higher T_g of amylopectin (~500 K) compared that of isotactic polystyrene and nylon-6, with 363.5 K, and 303 K, respectively.[9]

Equations 3 and 4 describe the effect of water content on T_g and T_m. The linear relationships listed in Table 1 relate log G_0, U*, and K_g to water content. This led to the simulation of effect of water content on rate of isothermal retrogradation of wms over a wide range of water contents (10 to 50% w.b) and storage temperatures (0 to 80°C). (Figure 5)

The predicted lines were tested against experimental isothermal recrystallization data, and a satisfactory agreement was found (Figure 6). This agreement is particularly rewarding, since most points of the 40°C data set were not used in the first part (temperature domain) of this analysis.

TABLE 1
Linear Dependence of the Lauritzen-Hoffman Parameters on Water Content (% dry solid basis). The values reported by Lauritzen and Hoffman[9] for isotactic polystyrene and nylon-6 are included for comparison.

	Slope	Intercept	Isotactic Polystyrene	Nylon 6
log G_0	–0.1044	5.75	N/A	N/A
U*	–81.4	8464	6427	5983
K_g	–3422	3.2 10^5	1.2 10^5	1.74 10^5

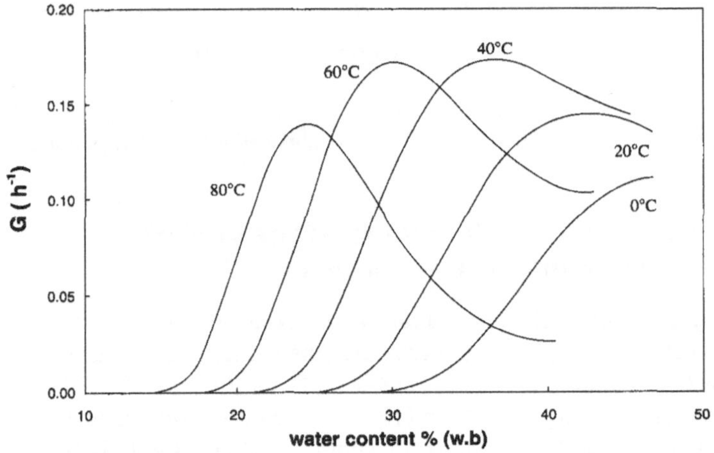

FIGURE 5 Lauritzen-Hoffman simulation of effect of storage temperature on rate of isothermal recrystallization of starch-water systems as a function of water content (adapted from Farhat et al.).[6]

The practical outcome of isothermal simulation is the contrast between the effect of increasing the water in the sample over the moisture content range of interest for this study (20 to 60% dry basis, i.e., 17 to 38% wet basis) for a sample stored at 40°C as compared to 80°C. While the increase of water content enhanced the rate of retrogradation at 40°C, the opposite behavior is observed at 80°C.

CONCLUSION

This chapter demonstrates the potential of the polymer science, or material science approach pioneered in the 1980s by various groups in the United States and the United Kingdom for understanding food systems behavior. The theory developed by Lauritzen and Hoffman for the crystallization kinetics of synthetic polymers was

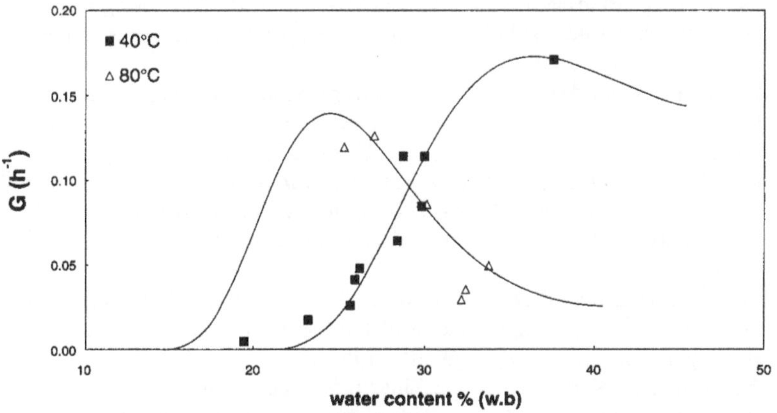

FIGURE 6 Comparison between predicted and measured isothermal retrogradation rates as a function of water content (adapted from Farhat et al.).[6]

applied successfully to starch retrogradation. This approach enabled a rational understanding of the effects of water content and storage conditions on the kinetics of starch retrogradation. Farhat et al.[22] offered a generalized hypothesis regarding the effect of sugars on the rate of starch retrogradation, an area where contradictory reports can be found in the literature.

It will be interesting to perform similar work with more frequent temperature domain data points over a range of water contents to investigate whether there is a change in kinetic parameters as retrogradation evolves from crystallization of amylopectin in the A-polymorph (at high crystallization temperatures and/or low water contents) to crystallization in the B-polymorph (at low crystallization temperatures and/or high water contents). Such a distinction was not detected in this study, probably due to the limited number of temperature domain data points, and maybe partly at the origin of the variation of K_g with water content.

REFERENCES

1. Katz, J. R., Gelatinization and retrogradation of starch in relation to the problem of bread staling, in *A Comprehensive Survey of Starch Chemistry*, Vol. 1, Walton, R. P., Ed., Chemical Catalog Company, New York, 1928, 100.
2. Hebeda, R. E. and Zobel H. F., *Baked Goods Freshness: Technology, Evaluation and Inhibition of Staling*, Marcel Dekker, New York, 1996.
3. Ring, S. G., Colonna, P., L'Anson K. J., Kalichevsky M. T., Miles M. J., Morris V. J., and Orford, P. D., The gelation and crystallization of amylopectin, *Carbohydr. Res.*, 162, 277, 1987.
4. Wilson, R. H., Goodfellow, B. J., Belton, P. S., Osborne, B. G., Oliver, G., and Russel, P. L., Comparison of Fourier-transform mid–infrared spectroscopy and near-infrared reflectance spectroscopy with differential scanning calorimetry for the study of the staling of bread, *J. Sci. Food Agric.*, 54, 471, 1991.
5. Avrami, M., Kinetics of phase change. II. Transformation-time relations for random distribution of nuclei, *J. Chem. Phys.*, 8, 212, 1940.

6. Farhat, I. A., Blanshard, J. M. V., and Mitchell, J. R., The retrogradation of waxy maize starch extrudates: effects of storage temperature and water content, *Biopolymers*, 53(5), 411, 2000.
7. Le Botlan, D. and Desbois, P., Starch retrogradation study in presence of sucrose by low-resolution nuclear magnetic resonance, *Cereal Chem.*, 72(2), 191, 1995.
8. Lauritzen, J. I., Jr. and Hoffman, J. D., Theory of formation of polymer crystals with folded chains in dilute solution, *J. Res. Nat. Bur. Stand. Sect. A*, 64, 73, 1960.
9. Lauritzen, J. I., Jr. and Hoffman, J. D., Extension of the theory of growth of chain-folded polymer crystals to large undercooling, *J. Appl. Phys.*, 1973, 44(10), 4340, 1973.
10. Marsh, R. D. L. and Blanshard, J. M. V., The application of polymer crystal-growth theory to the kinetics of formation of the β-amylose polymorph in a 50% wheat-starch gel, *Carbohydr. Polym.*, 9, 301, 1988.
11. ten Brinke, G., Karasz, F. E., and Ellis, T. S., Depression of glass-transition temperatures of polymer networks by diluents, *Macromolecules*, 16, 244, 1983.
12. Flory, P. J., *Principles of Polymer Chemistry*, Cornell University Press, Ithaca, New York, 1953, 541.
13. Kalichevsky, M. T., Jaroszkiewicz, E. M., and Blanshard, J. M. V., A study of the glass transition of amylopectin-sugar mixtures, *Polymer*, 34(2), 346, 1993.
14. Orford, P. D., Parker, R., and Ring, S. G., Aspects of the glass transition behaviour of mixtures of cabohydrates of low molecular weight, *Carbohydr. Res.*, 196, 11, 1990.
15. Roos, Y. H., *Phase Transitions in Foods*, Academic Press, San Diego, 1995.
16. Lelièvre, J., Starch gelatinization, *J. Appl. Polym. Sci.*, 18, 293, 1974.
17. Donovan, J. W., Phase transitions of the starch-water system, *Biopolymers*, 18, 263, 1979.
18. Biliaderis, C. G., Maurice, T. J., and Vose, J. R., Starch gelatinization phenomena studied by differential scanning calorimetry, *J. Food Sci.*, 45, 1669, 1980.
19. Whittam, M. A., Noel, T. R., and Ring, S. G., Melting and glass/rubber transition of starch polysaccharides, in *Food Polymers Gels and Colloids*, Dickinson, E., Ed., The Royal Society of Chemistry, 277, 1991.
20. Couchman, P. R. and Karasz, F. E., A classical thermodynamic discussion of the effect of composition on glass transition temperatures, *Macromolecules*, 11, 117, 1978.
21. Farhat, I. A. and Blanshard, J. M. V., On the extrapolation of the melting temperature of dry starch from starch-water data using the Flory-Huggins equation, *Carbohydr. Polym.*, 34(4), 263, 1997.
22. Farhat, I. A., Descamp, M., Blanshard, J. M. V., and Mitchell, J. R., The effect of sugars on the retrogradation of waxy maize starch — sugar extrudates, *Cereal Chem.*, 77(2), 202, 2000.

Index

E

F

G

H

I

K

L

M

R

S

T

U

V

W

X

Y

Z